MATHEMATICAL MODELS AS A TOOL FOR THE SOCIAL SCIENCES

CÉ

MATHEMATICAL MODELS AS A TOOL FOR THE SOCIAL SCIENCES

Edited by

Bruce J. West[†]

Formerly a Fellow of the Institute of Fundamental Studies,
Department of Physics and Astronomy,
University of Rochester, Rochester, New York 14627

[†]Presently with The La Jolla Institute, P.O. Box 1443,
La Jolla, California 92038

GORDON AND BREACH SCIENCE PUBLISHERS
NEW YORK ● LONDON ● PARIS

Copyright © 1980 by Gordon and Breach, Science Publishers, Inc.

Gordon and Breach, Science Publishers, Inc.
One Park Avenue
New York, NY 10016

Gordon and Breach Science Publishers Ltd.
42 William IV Street
London WC2N 4DE, England

Gordon & Breach
7-9 rue Emile Dubois
75014 Paris, France

Library of Congress Cataloging in Publication Data

Main entry under title:

Mathematical models as a tool for social sciences.

1. Social sciences—Mathematical models—Addresses,
essays, lectures. I. West, Bruce J.
H61.M427   300'.1'51   79-23974

Dedicated to my wife Sharon and sons Jason and Damien

TABLE OF CONTENTS

ACKNOWLEDGMENT

I would like to thank Professor Elliott Montroll for
his friendship, for making available the facilities
of the Institute of Fundamental Studies and for his
unflagging interest in a spectrum of problems.

INTRODUCTION

Bruce J. West*

La Jolla Institute
P.O. Box 1443
La Jolla, California 92038

This book is an outgrowth of a joint physics-mathematics seminar, organized by W. Badger, C. E. Watts, and B. J. West and presented at the University of Rochester. The intent of the seminar was to apply the techniques of mathematical modeling, which have proven themselves of such value in the physical sciences, to problems in the social sciences. The audience for this seminar was expected to be a composite of all levels of the University community. Therefore, the format was to present individual solutions to particular problems in the social sciences and in so doing reveal the motivation behind the mathematical techniques used. Very often it is not the modeling of particular problems which cause dissatisfaction among social scientists, but rather the obscure way in which the models evolve. It is hoped that the mosaic of mathematical methods presented in the following lectures will provide insight into all stages of model building, including inception.

The level of mathematical complexity varies markedly from lecture to lecture. This is due in part to the divergent backgrounds of the individual authors; only three of the nine lectures being given in the "specialty" of the individual author. This willingness of people to step outside the sanctuary of their field of specialization leads to fresh views and interesting developments in the treatment of problems. The lectures, therefore, are not intended to have the finished appearance of journal publications, but more the exploratory feeling of a report where methodology rather than solutions of particular problems are of central importance.

There is no one way, and indeed no best way, to construct a mathematical model of a natural system. One could, for example, study a system until all the major characteristics have been determined; then associate variables with each of these characteristics. Relationships between these variables may be hypothesized to reproduce the dynamics of the system. This approach to modeling is called simulation and is generally carried out on a computer. A more analytic method might be to select one or two of the more important variables (determined in some manner exterior to the system) from the system and treat the remaining variables in an approximate manner. This might be done by perturbing the privileged variables in a random manner, resulting in a stochastic model of the

*Formerly with the Institute of Fundamental Studies, Department of Physics and Astronomy, University of Rochester, Rochester, N.Y. 14627.

system. Alternatively, the remaining variables might be deterministic perturbations on the two-variable model which would lead to small corrections or changes. A third approach might be to select a structural model of sufficient mathematical generality so as to be applicable to many systems of a given *kind*. Then one may search for a natural system to which this general model may be adapted. Each of these methods of model building, among others, is used in the following lectures.

The philosophy behind these lectures, at least that of the Editor, has been that any problem which may be well formulated verbally, may be well formulated mathematically. Unfortunately, this is not to say that a problem so formulated will be amenable to a solution. It is often the case that the complexity of the problems which arise in the social sciences elude solution by presently known methods. This limitation in the availability of mathematical techniques constrains us to the position of constructing models. The implicit assumption in the construction of a model is that one has isolated the main characteristics of a complex event, and has been able to organize them into a rationally coherent structure. In general a model is a simplified replica of the system* we wish to understand. Understanding the model provides insight into the phenomenon being investigated without the restriction of being comprehensive.

(1) Professor R. W. Fogel takes the position in the first lecture that even in a discipline as traditionally literary as history, there is an implicit dependency on mathematics. He argues that a great many of the interpretations of historical situations and events have been, and will continue to be, changed when these implicit mathematical assumptions have been made explicit and their consequences explored. The historical fiction that slavery in the ante-bellum south was unprofitable is dispelled by way of example in the use of the new methodologies. Although the formal mathematical context of this lecture is light, it serves as an excellent introduction to the new attitude toward the use of mathematics in an area of research where its use has long been neglected.

(2) In a spirit similar to that of the preceding lecture, Professor A. O. Dick attacks the problem of serial learning in pyschology. The wealth of experimental data, as well as the limited success of previous models provides a framework in which a fresh viewpoint leads to a modest but discerning model. The model uses the notions of sets and set multiplications in conjunction with artificial intelligence analogies to predict error curves encountered in serial learning experiments. The model both explains the results of Professor Dick's experiments and others, and also poses questions which would not otherwise have been asked.

(3) Professor J. Keilson's lecture departs from those preceding it in that he treats a general *type* of situation as opposed to a particular problem. He treats the general problem of a contractor who wishes to select jobs from a series of job offers over a long period of time so as to maximize his average income per average interval of time worked. This lecture typifies the method of model construction where one fabricates an idealized environment for the process considered; constructs the model within this idealized environment and then solves the model problem. Both the strengths and weaknesses of this approach are discussed in the lecture.

(4) The lecture by A. Budgor and B. J. West is again an example of a model constructed from a plethora of data. The authors discuss the floods and droughts in the Nile River Valley, using 1300 years of well-established data. The model which is developed is stochastic in nature

---

*The terms system, event, phenomenon, etc., will all be used to describe the problem under investigation.

and leads to the well-known Gumbel distribution for extreme events. The intent of the lecture was to test the applicability of extreme value theory to yearly river maxima and minima and in particular to those of the Nile. The lecture discussed how a general theory (such as extreme value theory) may be modified so as to be made applicable to a particular problem.

(5) Professor Kemperman's lecture exemplifies the process of refinement which models undergo from their simplest beginning to their subsequent generality. He investigages the transmission of genetically controlled physical traits from generation to generation under different systems of mating. The investigation assumes that equilibrium or unchanging genotypic distributions exist for large populations. The form of these distributions is sought under different mating systems. For example, the effect of taboos, i.e., mating restriction of either a social or evolutionary origin, on the equilibrium distribution is explored. This latter effect is applied to the mating of a certain species of pigeons and the theory is found to explain the observed distribution.

(6) Professor W. Riker's lecture is a prototype of the axiomatic method of model construction. He constructs a theory of political coalitions based on the elements of n-person game theory. The necessary assumptions about the behavior of men to ensure the applicability of game theory to this problem are both reasonable and direct. The implications of the model, however, are both profound and somewhat disturbing. The major result of the model, which is the size principle, states that: coalitions grow to (shrink to) the minimum size necessary to win and no larger. This theory would then explain why coalitions such as the League of Nations and the United Nations are generally ineffective.

(7) The phenomenon of speculation has for a long time intrigued both the gambler and businessman alike. B. J. West explores this notion by attempting to isolate those characteristics of speculation which are present in a market situation. The model is normative and stochastic in nature since the characteristics of large groups are being discussed.

(8)-(9) W. Badger discusses the inequity in the distribution of wealth which is a traditional problem of concern to the economist, sociologist and political scientist. The methods employed were developed in information theory and lead to a least biased form of the distribution of wealth, while using all the information one has available. His investigation results in a distribution which fits wealth data over its entire range. Further, the two previous distributions, i.e., Praeto and lognormal, which were used at the extremes of the wealth distribution may be obtained from Badger's distribution by taking the appropriate asymptotic limits.

HISTORIOGRAPHY AND RETROSPECTIVE ECONOMETRICS*

Robert William Fogel

University of Rochester

*Copyright © 1970 by Wesleyan University. Re-
printed from *History and Theory*, Volume IX, No. 3,
by permission of Wesleyan University Press.

I. THE DOMAIN OF ECONOMETRIC HISTORY

My assigned topic is "Historiography and Retrospective Econometrics."
I want to address myself to only a limited aspect of the literature sug-
gested by that title, and I shall confine my remarks to the work of
Americans who have been writing on the history of their economy.

The systematic and self-conscious application of mathematical
methods to the study of American economic history is now some ten years
old. I stress the words "systematic" and "self-conscious" because it is
very easy to find earlier examples of one or another writer, on one or
another topic, at one or another time, who deliberately applied mathematics
to historical problems. And as I shall argue shortly, unconscious or subliminal applications are as old as the discipline. I am interested therefore in the systematic, self-conscious use of these methods.

A decade is not a very long time. During this short span econometric history has changed from a novelty into the predominant form of
research in American economic history. I say this with full knowledge
that the methodology of the new economic history is still highly controversial. However, I believe that these methodological debates are essentially rear guard actions.

The predominance of the new type of research is evidenced by a number of things. It is, for example, evidenced by the number of articles
published by American journals which are econometric in their basic character. On another occasion, I pointed out that one-third of the then
current issue of the *Journal of Economic History* (1966) employed quantitative methods. In a later issue of the *Journal* (1968) half of the articles
apply these methods. If I extrapolated naively, I would come to the conclusion that 100 per cent of the articles would be econometric in six
years and seven days.

The predominance of the new approach is also demonstrated by the
fact that the main postgraduate centers of economic history in the United
States, those departments which produce most of the Ph.D.'s in economic
history, now routinely teach econometric history. I include in this
class such institutions as Harvard, M.I.T., Yale, Columbia, Pennsylvania,
Johns Hopkins, Chicago, Purdue, Northwestern, Wisconsin, the University
of California at Berkeley, Stanford, and the University of Washington.

Thirdly, this predominance is demonstrated by the great interest displayed in the findings of the new work. The various essays produced by econometric historians have not gone unnoticed. Quite the contrary, virtually every new piece of econometric history has created some sort of controversy. To say this is not to imply the validity of their findings. I merely wish to stress that these findings have been of substantial interest to important segments of the profession.

Still another index is the number of articles that have been written on the methodology of the new work.[1] At last count there were over thirty such articles, each of which has sought to define what was unique or not unique about the new economic history, what was good or bad about it.

It is also worth noting that the emergence of econometric history coincides with a significant change in the general atmosphere of the discipline. When I first started going to meetings of the Economic History Association in the late 1950's, I was surprised to find many of the older participants referring to a renaissance in economic history. Now we all know that the renaissance came after the dark age. To someone just about to enter the profession, the discovery of this sense of recently dispelled pessimism was more than a little upsetting. The pessimism was due partly to the fact that very few young people had been coming into the field. The age gap was quite evident at the meetings. Many of the participants were in their fifties and sixties and there were some younger people in their twenties and early thirties. But there was virtually no middle. Well, this situation has changed. There has been a marked and steady influx of young people into the discipline over the past decade. These years have also been marked by a series of stimulating controversies of high intellectual caliber. The current atmosphere is dominated by a feeling of vigor in the discipline. I attribute much of the change in atmosphere to the new work.

The introduction of mathematical methods has not resulted in a narrowing of the range of issues treated by the discipline. I stress this point because it is frequently asserted that mathematical methods are only of very limited usefulness in economic history, that there is only a narrow range of propositions--and usually not the most important propositions--to which mathematical methods may be applied. I do not mean to suggest that every issue in economic history lends itself to mathematical methods, nor that all issues which in principle lend themselves to mathematical methods can in fact be dealt with adequately. Some historical questions involve relationships that go beyond the set currently covered by mathematics. In other cases the equation systems required to describe a given reality may be insoluble. Or it may be that, although one can define a model and solve it, the date required to estimate the parameters of the model are not available.

However, these limitations have not forced practitioners of the new economic history onto the sideroads of their discipline. Quite the contrary, the topics on which they have focused are the classical issues of American economic history. The first big concentration of research was on the analysis of the slave economy of the ante-bellum South. Econometric historians have examined such central problems as the profitability of slavery, the economic viability of the slave system, the effect of slavery on the course of Southern economic growth, and the relative efficiency of the salve system of labor organization.[2] A second focus of research has been the developmental impact of transportation improvements. This question, which is so closely intertwined with the frontier hypothesis, has been one of the great concerns not only of economic but of social and political historians.[3] Other prominent areas of investigation by econometric historians include the analysis of the determinants of urbanization, the influence of foreign trade on the creation of a market

economy, the explanation for the growth of manufacturing industries, the
measurement of the burden of the British imperial system on the American
colonies, the efficiency of federal land policy, and the factors affecting the diffusion of new innovations. I submit that this list contains
no problem which American historians would brand as trivial, as a mere
showcase for mathematical sleight of hand.[4]

To stress that the focus of econometric history has been on classical subjects is not to imply that the new work has only decorated old
boxes with elaborate ribbons. I would find little merit to methods which
were elegant but told us nothing new. Indeed, we should expect the com-
mittees which award research grants to demand more than pretty frills.
While some of the new work has merely served to confirm conclusions
reached long ago, most of these investigations have resulted in substantial challenges to traditional interpretations.

Take the issue of slavery. Econometric research has largely revised the once standard view that the slave economy of the ante-bellum
South was unprofitable, stagnant, inefficient, and moribund. On the
basis of better data and more refined analytical techniques than were at
the disposal of earlier scholars, the new economic historians have demonstrated that slavery was quite profitable. To put it in contemporary
terms, as an investment opportunity, slavery was the growth stock of the
1850's. Thus when slaveowners invested in slaves, it was not because
they were doddering idiots wedded to an economically moribund institution.
Nor was it because they were noble men who were sacrificing their personal
economic interests to save the country from the threat of barbarism.
Perhaps slaveowners were nobly motivated. If so, they were well rewarded
for their nobility--with average rates of return in tne neighborhood of
10 or 12 per cent per annum. New measurements also indicate that the
slave economy was growing between 1840 and 1860. Far from being the lag-
gard region, the rate of growth of per capital income in the South ex-
ceeded the national average. But perhaps the most startling of the new
findings is the discovery that Southern agriculture was nearly 40 per
cent more efficient in the utilization of its productive resources than
was Northern agriculture.[5]

Consider the case of cotton textiles. Textbook chapters on the
growth of cloth production during the ante-bellum era typically center
on the mechanical inventions that revolutionized the industry. Such discussions frequently begin by citing data on the increase in the consump-
tion of raw cotton, or on the change in the number of spindles in operation, between 1810 and 1860. The rapid growth indicated by these measures
is usually attributed to development of the factory system. The rise of
the factory is in turn made to depend on a series of mechanical inven-
tions--the flying shuttle, the mule, the water frame, the power loom.
The remainder of the chapter will be devoted to detailed verbal descrip-
tions of these machines (pictures are not usually supplied) which are
difficult, if not impossible, to visualize. The impression left in the
reader's mind, even if it is not explicitly stated by the author, is that
the growth of cotton textiles was primarily due to mechanical inventions.

In a recent paper, Robert B. Zevin set out to determine how much of
the growth in the American cotton textile industry between 1815 and 1833
could be attributed to mechanical innovations and how much was due to
other factors.[6] The "other factors" are increases in demand, reductions
in the cost of inputs, and plant expansion. There are a number of sur-
prises in Zevin's essay. He finds that 55 per cent of the growth rate
over this eighteen-year period was due not to shifts in supply but to a
rapid rise in the demand for cotton cloth. Even if there had not been a
single new invention, even if the old technology had persisted without
modification, the rise in demand for cotton cloth would have caused the

industry to expand at a lusty rate. Moreover, not all of the remaining 45 per cent increase that Zevin assigns to supply shifts was due to technological change. Reductions in the cost of inputs, particularly in the price of raw cotton, accounted for about 10 per cent of the expansion in output. Zevin assigns the credit for the balance of the observed growth to new machinery and the elimination of a market disequilibrium which existed at the beginning of the period. The disequilibrium was due to the fact that demand, stimulated by the embargo on British cotton during the War of 1812, rose more rapidly than supply. The elimination of the gap turned on the expansion of plant capacity, an expansion which could easily have taken place under the old technology. Zevin did not have the data required to estimate with confidence the effect on output of the elimination of the bottleneck in plant capacity. But one crude calculation indicated that it might have been substantial. Thus Zevin's essay suggests that the new mechanical inventions, which have received so much of the attention of historians, did not account for more than 35 per cent of the growth of the cotton textile industry; and they may have accounted for even less.

One of the livelier debates on government economic policy during the nineteenth century centers on the effect of the tariff on the growth of the iron industry between 1842 and 1860. Resolution of the dispute rests on the responsiveness of the demand for domestic iron to changes in the price of imported iron. That responsiveness is measured by a coefficient called "the cross-elasticity of demand." While neither the free traders nor the protectionists ever provided evidence on the size of this coefficient, the persuasive arguments of F. W. Taussig[7] made the free-trade view predominant for more than three-quarters of a century. In a study that Stanley Engerman and I recently completed,[8] we used econometric methods to estimate the demand curve for American iron. Our results show that the crucial cross-elasticity is quite large. A large coefficient means that the tariff was an important factor in determining the level of domestic iron production during the ante-bellum era.

I could greatly extend the list of new findings. But the examples cited are sufficient to demonstrate that the methodology of the new economic history cannot be ignored. For the substantive findings of the new work are quite different from the findings of the older research. And the new results are closely connected to the methods used to produce them. Let me therefore turn to the consideration of some basic features of methodology of econometric history.

II. METHODOLOGY

The hallmark of the new economic history is its emphasis on measurement and its recognition of an intimate relationship between measurement and theory. I do not mean to suggest that cliometricians introduced quantification into economic history. But they have pushed measurement in directions which are substantially different from those previously pursued. In the past, quantification was limited largely to the collection and presentation of data in their original form or to relatively simple reclassifications of original data. With the exception of the extensive work on the construction of price indexes, little was done to recombine primary data into constructs which conformed to rigorously defined economic concepts. It is such operations on primary data which constitute the central interest of the econometric historians.

The new quantitative work may be divided into three categories. First, economic theory and mathematical models have been used to reconstruct the dimensions of economic institutions or processes from surviving fragments of data. In this respect econometric historians may be

compared with paleontologists who attempt to reconstruct prehistoric animals on the basis of skeleton fragments and biological theory. Second,
primary data have been recombined into constructs which conform to rigorously defined economic concepts. National income statistics are perhaps
the best-known of these constructs. Measures such as Gross National
Product are not set forth in historical documents. One has to sift
through census schedules and the surviving records of primary business
units to gather the data from which that measure of aggregate output may
be compounded. Third, much effort has been devoted to measuring that
which can only be measured indirectly. For example, a central issue in
determining the social saving of railroads is the measurement of the cost
of the time lost in shipping by a slow medium of transportation. This
cost is not recorded in profit or loss statements or in any of the other
normal sources of business information. Consequently it must be measured
indirectly through a method which links the desired information to data
which are available. In my book on American railroads[9] I argued that the
problem could be resolved on the basis of the nexus between time and inventories. Inventories are necessary to bridge the time-gap required to
transport a commodity from its source to its destination. Hence any increase in transportation time could be compensated for by increasing the
size of inventories in such a way as to maintain exactly the same temporal
flow of commodities as was possible with rapid transportation. The cost
of slow transportation, then, is the cost of maintaining that extra
inventory.

The list of tools available to the econometric historian concerned
with these new problems of measurement is much larger than that which was
available in the past. Cliometricians are the beneficiaries of the advances made over the past several decades in statistics and other applied
fields of mathematics. While regression analysis is the most frequently
used tool of econometric historians, it is by no means the only one.
Other mathematical models which have been applied include linear program-
ming, the hypergeometric distribution, input-output analysis, simulation
models, analysis of variance, Von Neuman-Morgenstern utility indexes, factor analysis, and Markov chains.

From the foregoing, it is clear that the new techniques of measurement depend heavily on theory. This intimate union is most obvious in
the case of indirect measurement. For indirect measurement presupposes a
functional relationship between that which you desire to measure and that
which you are able to measure. Indirect measurement can only be carried
out if the functional relationship, or at least certain of its characteristics, are known.[10]

The intimate union between measurement and theory is also obvious
when economic historians attempt to assess the net benefit of particular
policies, institutions, or processes. For example, the net benefit of
railroads is measured by the difference between the level of national income that was actually achieved and the level that could have been
achieved in the absence of railroads. Of course, the only economy we can
observe is the one that actually existed. We cannot observe the economy
that would have existed in the absence of railroads. Hence we can make
statements about the effect of the absence of railroads on the level of
national income only if certain characteristics of the economy can be described by functional relationships, functions which permit us to predict
what the values of relevant variables would have been in an economy
without railroads.

III. THE ROLE OF MATHEMATICS

To many commentators the most distinctive feature of the new economic history is its introduction of mathematics (equations) into what
used to be a literary discipline. That view gives too much credit to the
cliometricians. Mathematics has always been present in economic history.
In the past, however, it was implicit, covert, and subliminal. The contribution of cliometricians is that they have made implicit mathematics
explicit. Moreover they are concerned with whether the implicit equations crucial to some arguments are the correct equations. Does the particular functional form implicitly designated by an historian conform to
the reality that he wishes to describe?

Permit me to cite two examples which illustrate both the prevalence
and treacherousness of covert mathematics. The first example is drawn
from the literature on the early industrialization of the American economy.
Iron and steel is one of the most frequently discussed manufacturing sec-
tors. Textbook writers, and others, note that while the output of this
industry grew at only a moderate rate before 1840, it expanded quite
rapidly between 1840 and 1860. To support this statement it is common to
cite the following figures on pig iron production:[11]

<u>Tons of pig iron</u>

1840	287,000
1860	988,000

But pig iron is only one of the products of the iron industry, not
its total output. As measured by value-added, pig iron accounted for only
about one-sixth of the total output of the industry in 1860.[12] Consequently historians who use the rate of growth of pig iron production to
measure the rate of growth in total output are implicitly assuming the
following functional relationship:

$$T = mP \qquad\qquad (1)$$

where T = total output
P = pig iron production
m = a constant.
In other words, they are implicitly stating that total output of the industry is directly proportional to the output of pig iron. This is the
simplest of all equations. It is a linear equation with a zero intercept.

Just because equation (1) is simple does not necessarily mean that
it distorts reality. As a matter of fact, equation (1) is a good description of the relationship between pig iron and total production from
1830 to 1860. However, it is a poor description of the relationship
between these variables over the period from 1860 to 1900. Thus if one
uses pig iron as a measure one will conclude that the iron industry stag-
nated between 1860 and 1870, although an index of total output shows that
the Civil War decade represented an era of unusually rapid expansion.[13]
Since the growth of the iron industry has been made an issue in the debate over the effect of the Civil War on Northern industrialization, the
inapplicability of equation (1) for the period after 1860 is not an empty
issue for historians.

The second example is drawn from the literature on American Negro
slavery. U. B. Phillips published his famous essay on *The Economic Cost
of Slaveholding in the Cotton Belt* in 1905.[14]   In this paper he argued
that slaves were an unprofitable investment. To support his contention,
Phillips assembled time series on the prices of slaves and raw cotton.
These series showed that from 1815 on, slave prices rose more rapidly than

cotton prices. According to Phillips that fact was sufficient to establish the proposition that the profitability of slavery must have declined over the period. Indeed, since the ratio of slave to cotton prices was much higher in 1860 than it had been in 1815, he drew the conclusion that by the eve of the Civil War, slavery had become unprofitable.

Equation (2) is the algebraic representation of the Phillips argument:

$$\overset{*}{i} = k(\overset{*}{P}_C - \overset{*}{P}_S)\tag{2}$$

where i = the rate of return
      k = a constant
      P_C = the price of cotton
      P_S = the price of slaves
      * = an asterisk over a variable stands for the rate of change in the uncapped variable; thus $\overset{*}{i}$ is the rate of change in i.

Equation (2) states that the rate of change in the rate of return is directly proportional to the difference between the rates of change in cotton and slave prices. When $\overset{*}{P}_S$ is greater than $\overset{*}{P}_C$, not only will $(P_C - \overset{*}{P}_S)$ be negative but, if equation (2) is correct, $\overset{*}{i}$ will also be negative. As in the previous example, the interpretation of one of the major issues of American history turns on the implicit assumption that certain variables are related to each other by a linear equation with a zero intercept.

I do not mean to give the impression that Phillips was naive. Quite the contrary, the issues in the economics of slavery which have occupied so much of the attention of econometric historians during the past decade are those which he defined. It is surprising that Phillips knew as much about capital theory as he did. Not only is equation (2) related to the equation which Alfred H. Conrad and John R. Meyer subsequently used to estimate the rate of return on an investment in slaves, but in one of the footnotes of *American Negro Slavery*, Phillips explicitly referred to the basic form of their equation,[15] namely:

$$P_S = \frac{H}{i}\left[1 - \frac{1}{(1 + i)^n}\right]\tag{3}$$

where P_S = the price of a slave
      H = the expected average annual net income to be earned from the employment of the slave
      n = the expected number of years between the purchase of the slave and his death
      i = the rate of return on the purchase price of the slave.

Equation (3) states that the price an investor was willing to pay for a slave was equal to the discounted present-value of the average annual net income he expected to earn as a result of owning the slave. This is the standard formula for capitalizing an income stream. It is the equation used for determining the price of a long-term annuity or, with slight modification, the price of a long-term bond. Phillips was not only aware of the similarity between investment in a slave and in a long-term security such as a bond, but built much of his argument on that similarity.

It can be shown that the expression for the change in the rate of profit implied by equation (3) is not equation (2) but equation (4):

$$\overset{*}{i} = k[\phi\overset{*}{P}_C - \overset{*}{P}_S) + \phi(\overset{*}{Q} - \overset{*}{L}) + (1 - \phi)\overset{*}{M}].\tag{4}$$

All of the symbols of equation (4) have been defined previously except

$(\overset{*}{Q} - \overset{*}{L})$ = the rate of change in the productivity of slaves--
      in output (Q) per slave worker (L)

$\overset{*}{M}$ = the rate of change in the cost of slave maintenance

ϕ = the ratio of the gross annual income (H_g) earned on
   a slave to the net annual income (H); i.e., H =
   H_g - M and ϕ = H_g/H

k = a constant close to 1; its exact magnitude depends
   on the base-period values of i and n.

A comparison between equations (2) and (4) reveals that equation
(2) is merely a special case of equation (4). Equation (4) will reduce
to equation (2) when ϕ = 1 and $\overset{*}{Q} - \overset{*}{L}$ = 0. Consequently the evaluation of
the Phillips thesis comes down to the question of whether Phillips was
justified in implicitly assuming that ϕ = 1 and $\overset{*}{Q} - \overset{*}{L}$ = 0. For only then
is information on the change in the ratio of cotton to slave prices *alone*
sufficient to determine that profits were declining. Clearly the assumption that ϕ = 1 is false. Since ϕ = H_g/H, the condition for ϕ to be equal
to one is that expenditures on the maintenance of slaves (M) were zero--
that slaves were being starved to death.[16] All available evidence, including the evidence assembled by Phillips on the material conditions of
slave life, refutes this assumption. Available evidence also contradicts
the assumption that there was no increase in the productivity of slaves.
Indeed a recent estimate based on a sample of data from the manuscript
schedules of the census suggests that the rate of growth in salve productivity ($\overset{*}{Q} - \overset{*}{L}$) between 1840 and 1860 may have been as high as 2.9 per
cent per annum.[17]

Thus the thesis that slavery was unprofitable, a proposition that
dominated the historiography on the ante-bellum South for a half century,
was based on a false equation. The relationship between the rate of
change in profit and the rate of change in the ratio of slave to cotton
prices was not described by a linear equation with a zero intercept. The
correct functional form was more complex than Phillips realized.

The lesson of these examples is simple. The prohibition of explicit
equations will not eliminate mathematics from historiography. It will
merely impede the effort to determine whether the implicit equations em-
bedded in important arguments are true or false, whether these equations
are accurate depictions of the reality with which historians are concerned.

IV. THE PROBLEMS OF COUNTERFACTUALS

To some critics, the most alarming feature of econometric history
is the prominence given to counterfactual conditional statements. Essays
embodying such statements have been called "quasi-historical," "ficti-
tious," and speculative intrusions into an empirical discipline.[18] These
characterizations reflect a misunderstanding of the nature of counterfactuals and of their central role in historiography. I attribute much
of the confusion to the cavalier treatment in some philosophical discussions of the array of issues posed by counterfactuals. Among the worst
villains are certain physical scientists who, perhaps because they never
had to confront the issue in their practical work, take the position that
counterfactual conditional statements cannot be verified and, hence, fall
beyond the pale of science.

Unfortunately, historians cannot indulge in the luxury of dismissing
the issue. Like it or not, counterfactual conditional statements are too
integral a feature of their discourse to be banished. Should we advocate

that historians give up the practice of making judgments about mistakes? Do we mean to exclude from history such statements as: "Woodrow Wilson miscalculated the consequences of his failure to appoint a prominent Republican to the delegation that represented America at the Paris peace conference" or "Andrew Johnson played into the hands of his enemies by suspending Stanton and making Grant *ad interim* Secretary of War?" There are few historians who would accept a ban on such judgments. To do so would transform history into mere chronology.

But historians who write that Wilson or Johnson made mistakes are implicitly asserting that the particular course of action that each man followed was inferior to an alternative that was available to him. In branding actual behavior as an error, one presumes knowledge of the course of events in situations that never occurred.

If historians are not prepared to expurgate judgments about "what might have been" from their narratives, the real issue regarding counterfactual conditional statements is not whether we should make them, but how to establish criteria which enable one to determine the validity of such statements. This task is not as complicated as it has been made to seem by those who insist on forcing the discussion of counterfactuals to turn on paradoxical, but not very relevant, examples. Indeed, if we confine our attention to the type of counterfactual statements which are typically made in economic history, it will become apparent that the verification of these statements poses no special epistemological problems. Let me illustrate this contention.

Economic historians, as I have already argued, proceed on the implicit or explicit assumption that economic behavior may be described by functional relationships. Thus discussions of the factors responsible for the growth of the American iron industry have been cast in terms of demand and supply. Let us suppose[19] that these demand and supply curves are described by equations (5) and (6) respectively

$$Q = [bI^{\psi}P_i^{\varepsilon_i}]P^{-\varepsilon} \tag{5}$$

$$Q = (mA^{1/\alpha_3} W^{-\alpha_1/\alpha_3} r^{-\alpha_2/\alpha_3} K)P^{\gamma} . \tag{6}$$

The bracketed variables are the shift terms of the demand and supply curves, and

Q = the output of domestic iron
I = gross domestic investment
P_i = the price of imported iron
P = the price of Q
A = index of productive efficiency
W = wages
r = the cost of a unit of raw materials
K = input of capital
ψ = the investment elasticity of demand
ε_i = the cross-elasticity of demand with respect to P_i
ε = elasticity of demand
α_j = the output elasticities of the inputs; j = 1,2,3
γ = elasticity of supply
b,m = constants

Solving for the equilibrium value of Q we obtain

$$Q = (bI^{\psi}P_i^{\varepsilon_i})^{\gamma/(\gamma+\varepsilon)}(mA^{1/\alpha_3} W_1^{-(\alpha_1/\alpha_3)} r_1^{-(\alpha_2/\alpha_3)} K)^{\varepsilon/(\gamma+\varepsilon)}. \tag{7}$$

The rate of growth transformation of equation (7) is

$$\overset{*}{Q} = \frac{\gamma}{\gamma+\varepsilon}\left(\psi\overset{*}{I} + \varepsilon_i\overset{*}{P_i}\right) + \frac{\varepsilon}{\gamma+\varepsilon}\left(\frac{1}{\alpha_3}\overset{*}{A} - \frac{\alpha_1}{\alpha_3}\overset{*}{W} - \frac{\alpha_2}{\alpha_3}\overset{*}{r} + \overset{*}{K}\right) \qquad (8)$$

Equation (8) states that the rate of growth of output is an arithmetic average of the rate of growth of the shift term of the demand curve and the shift term of the supply curve.

Let us now consider F. W. Taussig's contention that a change in the tariff of 1846 would have had little effect on the growth of the output of the iron industry.[20] A change in the tariff would, of course, have affected the level of domestic iron production through its impact on the price of imported iron (P_i). If the structure of the market for iron is in fact described by equations (5) and (6), equation (8) can be used to measure the effect of a specified change in the tariff on the rate of growth of output. From equation (8) it is apparent that Taussig's contention that a large change in the tariff (and hence in $\overset{*}{P_i}$)[21] would have had little effect on the level of output reduces to the proposition that $\varepsilon_i\gamma/(\gamma+\varepsilon)$ was small.

From the foregoing discussion it is obvious that the crucial step in the verification of a counterfactual statement is the determination of the empirical validity of the explicit or implicit equation (or set of equations) which is purported to describe a specified reality. Given such an equation, one can determine what the value of the dependent variable would have been, if a contrary-to-fact value of a particular independent variable had occurred, merely by inserting that counterfactual value into the equation.

I do not mean to suggest that the determination of the empirical validity of the specification of a particular equation or of the magnitudes of particular parameters is an easy task. Indeed, many of the debates touched off by the new economic history turn on the estimates of the value of such parameters or on the consequences of misspecification of particular equations.[22] For example, my estimate of the social saving due to railroads involves the assumption that the long-run marginal cost of water transportation was constant or declining--that the function which relates the cost of water transportation to the volume transported had a zero or negative first derivative.[23] The challenge raised to this assumption has led me into a much more extensive search for data bearing on the shape of the cost function than I originally undertook. Similarly, Stanley Lebergott has argued that I fell into error in basing my estimate of the social saving on the market prices of rail and non-rail transportation. He contends that the forces of competition would have made the prices of these two services equal. Lebergott goes on to say that the inequality of rail and non-rail transport prices in my calculation must therefore be due to errors in measuring these prices. "A positive figure for the 'social saving,'" he concludes, "then becomes a measure of estimating error."[24] In asserting that competition should equate the prices of rail and non-rail transportation services, Lebergott is implicitly assuming that these services are perfect substitutes for each other. For then, and only then, would competition necessarily lead to an equalization of the prices.[25] The degree of substitutibility between two services is measured by a coefficient called the "elasticity of substitution." Lebergott's challenge has thus led me to collect the data necessary to estimate the elasticity of substitution between rail and non-rail forms of transportation.

Difficult as the resolution of these questions may be, they present no new methodological issues. The methodological points which have arisen are the standard ones discussed in textbooks of statistics and econometrics.

It is differences of opinion regarding the correct descriptive equations, not broad epistemological issues, which are at the root of the current debates regarding the validity of particular counterfactual conditional statements made by economic historians. For when the true descriptive equations are known, making counterfactual statements is equivalent to reading the values of the dependent variables which correspond to the specified values of the independent variables.

REFERENCES

1. Nine of these essays have been reprinted in Ralph L. Andreano, ed., *The New Economic History: Recent Papers on Methodology* (New York, 1970); others are cited there and in Robert W. Fogel, "The Specification Problem in Economic History," *Journal of Economic History* 27 (1967), 283-308; Robert W. Fogel and Stanley L. Engerman, *The Reinterpretation of American Economic History* (New York, Harper and Row, 1971); Maurice Levy-Leboyer, "La 'New Economic History'," *Annales* 24 (1969), 1035-1069.

2. Alfred H. Conrad and John R. Meyer, "The Economics of Slavery in the Ante-Bellum South," *Journal of Political Economy* 66 (1958), 95-130; Robert W. Fogel and Stanley L. Engerman, "The Economics of Slavery," Report 6803, Center for Mathematical Studies in Business and Economics, University of Chicago (January, 1968); and "The Relative Efficiency of Slavery: A Comparison of Southern and Northern Agriculture in the United States during 1860," (unpublished paper prepared for presentation at the Fifth International Congress of Economic History, August, 1970); Robert Evans, Jr., "The Economics of American Negro Slavery," in National Bureau of Economic Research, *Aspects of Labor Economics* (Princeton, N.J., 1962), 185-243; Yasukichi Yasuba, "The Profitability and Viability of Plantation Slavery in the United States," *Economic Studies Quarterly* 12 (1961), 60-67; Gavin Wright, *The Economics of Cotton in the Ante-bellum South* (unpublished Ph.D. thesis, Yale University, 1969).

3. Robert W. Fogel, *Railroads and American Economic Growth: Essays in Econometric History* (Baltimore, Md., 1964); Albert Fishlow, *American Railroads and the Transformation of the Ante-bellum Economy* (Cambridge, Mass., 1965).

4. For a more detailed review of the substantive findings of the new economic history see Andreano, *New Economic History*; Lance E. Davis, "'And It Will Never Be Literature': The New Economic History: A Critique," *Explorations in Entrepreneurial History* 6 (1968), 75-92; Maurice Levy-Leboyer, "La 'New Economic History'"; Gavin Wright, "Studies in Econometric History," (unpublished paper, Yale University, 1969). Some of the best-known essays in econometric history are included in Fogel and Engerman, *Reinterpretation of American Economic History*.

5. Fogel and Engerman, "The Relative Efficiency of Slavery." The debate on the economics of slavery is reviewed in Conrad and Meyer, "The Economics of Slavery in the Ante-Bellum South," and Fogel and Engerman, "The Economics of Slavery."

6. Robert B. Zevin, "The Growth of Cotton Textile Production After 1815" in Fogel and Engerman, *Reinterpretation of American Economic History*.

7. F. W. Taussig, "The Tariff, 1830-1860," *Quarterly Journal of Economics* 2 (1888), 314-346.

8. Robert W. Fogel and Stanley L. Engerman, "A Model for the Explanation of Industrial Expansion During the Nineteenth Century: With an Application to the American Iron Industry," *Journal of Political Economy* 77 (1969), 306-328.

9.   Fogel, *Railroads and American Economic Growth*.

10. It may, for example, be sufficient merely to know that the function is monotonic over some range.

11. The data, which are from the Censuses of Manufacturing for 1840 and 1860, are cited in James M. Swank, *History of the Manufacture of Iron in All Ages*, 2nd ed. (Philadelphia, 1892).

12. Fogel, *Railroads and American Economic Growth*, 256–257.

13. Fogel, "The Specification Problem in Economic History."

14. *Political Science Quarterly* 20 (1905), 257–275.

15. The equation is presented in chapters VI and VII of Arthur H. Gibson, *Human Economics* (London, 1909). In *American Negro Slavery* (Baton Rouge, La., 1966), Phillips cites Gibson and then states that his own discussion is "mostly in accord with Gibson's analysis" (359).

16. The reciprocal of ϕ is a measure of the exploitation of slaves (H/H_g). When $\phi = 1$, exploitation is complete. The owner expropriates 100 per cent of the income attributable to the slave's labor.

17. James P. Foust and Dale E. Swan, "Productivity of Ante–Bellum Slave Labor: A Micro Approach" (unpublished paper presented to the Workshop in Economic History, University of Chicago, February 24, 1967).

18. Andreano, *New Economic History*, 85–99; and Fritz Redlich, "New and Traditional Approaches to Economic History and Their Interdependence," *Journal of Economic History* 25 (1965), 480–495.

19. The model presented in equations (5)–(8) is discussed at length in Fogel and Engerman, "A Model for the Explanation of Industrial Expansion During the Nineteenth Century."

20. *Ibid.*, and Taussig, "The Tariff, 1830–1860."

21. Since $P_i = P_b(1+\lambda)$, $\overset{*}{P}_i = \overset{*}{P}_b + \frac{\lambda}{1+\lambda} \overset{*}{\lambda}$. Here P_b is the delivered price of imported iron before the tariff is levied and λ is the tariff rate.

22. Andreano, *New Economic History*, 151–175; Fogel, "The Specification Problem in Economic History"; Wright, "Studies in Econometric History."

23. Fogel, *Railroads and American Economic Growth*, 28.

24. Stanley Lebergott, "United States Transport Advance and Externalities," *Journal of Economic History* 26 (1966), 439.

25. Of course two imperfect substitutes could have the same price because the demand curve for their products happened to intersect their respective supply curves at identical prices. But such an unlikely accident is irrelevant in the context of Lebergott's argument.

A MATHEMATICAL MODEL OF SERIAL MEMORY

A. O. Dick

University of Rochester

One of the nice features about Psychology is that is is often easy to construct an immediate example of the phenomenon that is to be discussed. The example is usually clear and the point of it readily understood, although the mechanism which is operative during the example may not be understood at all. Let us then, by way of introduction, consider a simple experiment in Serial Learning. Let us suppose that an experimenter has in his possession a list of random characters; these characters might be letters, numbers, nonsense syllables, etc. These characters have been selected in some way that ensures their independence. For example, numbers may be selected by spinning a roulette wheel, letters may be randomly drawn from a list, or any other of a number of ways may be used to ensure the random nature of the characters. Next, the experimenter presents the characters to a subject, whether visually or auditorily, each in the same way in equally spaced time intervals throughout the list. The subject attempts to remember the entire list in the order presented and attempts to reproduce it in that order after the last character presented is presented.

This experiment is intended to analyze the mechanism of Serial Memory. One is able from this experiment to develop a general notion of the process which we intend to model. We will follow a historical path in our discussion both to understand this phenomenon and to see how previous models fell short of a complete explanation.

The experiments that were conducted prior to about 1960 were similar to the above example, but differed in that the lists were repeatedly presented to the subject until they were learned. In Figure 1 we present typical results from an experiment of this latter type.[1] In this figure the percentage of correct responses given by the subject is plotted against the position of the item in the list. The experiment is one in which a list of nonsense syllables is presented to a subject in an initial presentation. After the learning trial, the subject is again presented with the list and, as each character is presented to him, he must predict what the following character will be. Figure 1 then gives the results of such an experiment over a number of individuals. We note the following characteristics about the curve. First, that the characters at each end of the list are remembered more easily than those in the middle of the list. This can be interpreted to mean that they are more easily learned. Second, that the curve is slightly skewed to the right of center, indicating that the characters slightly past the middle of the list are the most difficult to learn. Third, that the curve has a generally smooth shape and a single minimum.

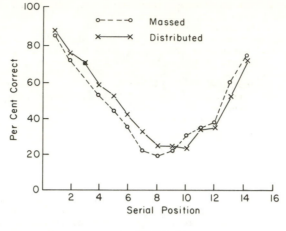

Figure 1

    If all empirical curves had the same general characteristics as the one in Figure 1, modeling the function would be rather simple. As is true with many aspects of behavior, the empirical values obtained vary as a function of the explicit instructions given to the subject on the implicit instructions (strategies) that a subject may employ. This point is illustrated in the next two figures.

    The results of an experiment similar to the first one mentioned above[2] are given in Figure 2 which was a one-trial experiment in which

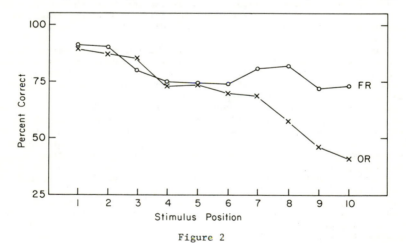

Figure 2

subjects were presented with a list of ten random letters. The subjects were either asked to recall the items in any order they wished (FR) or in the order presented (OR). We attempted to determine the dependence of serial memory on the rate at which the response from the subject was elicited. We found there was substantially no difference between the subjects responding at a fixed rate or at their own rate. The rate of output, then, can probably be ignored as an important determinant.

Each of the points in Figure 2 represents 240 independent experiments performed with University of Rochester students. We can see two distinct effects in the curves of Figure 2. In the free recall curve (FR), there is an improvement in recall just past the middle of the item list, i.e., the local maximum at item 8, in contrast with the single minimum of Figure 1. Further, in the ordered recall (OR) curve, there is almost an exponential decay in the learning curve as one goes down the list, i.e., no finite minimum.

We may represent the date of Figure 2 in a different form, as shown in Figure 3. Here we have scored the results of the experiments so that

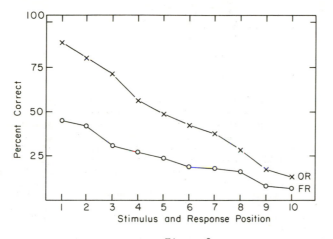

Figure 3

a person's recall of an item in the parallel stimulus and response position is registered as correct. We can see that the double minimum in the free recall curve (FR) of Figure 2 is no longer present. Similarly, the ordered response curve has smoothed out significantly so that if one were seeking a functional representation of the curve, an exponential would do nicely. (Both are smooth within boundaries of variance.) We are not, however, attempting to fit a particular curve in this discussion, but rather attempting to find a coherent model which will reproduce all the effects noted.

Historical Context. The bowed curve of Figure 1 is representative of the type of serial learning whose investigation was begun by Ebbinghaus[3] in 1885. Ebbinghaus believed that one formed associations in the learning of a list and that these associations facilitated recall. His experiments were conducted along the following lines. A list of length N would be learned, subject to some criterion, e.g., being able to repeat the list twice without error. The length of time necessary to learn this list would be recorded. Next, a second list would be derived from the first by a systematic interchange of letters. The time necessary to learn this second list, subject to the same criteria, would determine whether the associations formed in the first learning were disruptive or facilitory.

The idea of association formation was very strong in American psychology and, during the 1930's, was used to explain the bowed curve in Figure 1. C. L. Hull[4] used a model of associations similar to that shown

in Figure 4 where, for simplicity, we have used a six-item list. The notion was introduced that remote associations tend to inhibit immediate

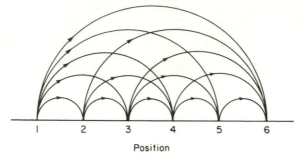

Position

Figure 4

associations, i.e., neighboring pairs, so as to impede learning. The
strength of the inhibition may be measured by the number of remote associations spanning the item of interest, i.e., the number of lines above
an item in Figure 4. This concept in conjunction with the assumption
that the number of repetitions of a list is a linear function of the
strength of the inhibitions yields a bowed curve similar to Figure 1.
The difficulty with this model is that the predicted curve is always symmetric about the middle of the list. The skewed character of Figure 1
is, therefore, not accounted for. Also, the assumption was made that
forward associations are stronger than backward ones; data do not support
this assumption. Both Ebbinghaus and Hull had assumed one particular
type of theory of learning.
       There are, in fact, two very different kinds of learning theory.
One is the single-trace model which assumes that the strength of memory
is proportional to the number of repetitions, e.g., suppose a single
black ball is maintained in an urn; subsequent black balls that enter
merge with the first, such that the *size* of the single ball changes.[5]
Although the analogy seems far-fetched, the process is possible in terms
of psychological phenomenon. The second type of theory assumes that each
repetition forms an independent trace so that, for a single learned sequence, there are multiple traces. The number of traces is then proportional to the strength of the memory, e.g., putting black balls in an
urn with other balls, where the strength of memory is proportional to the
number of black balls in the urn. With the neurophysiological and bio-
chemical information we have about neurons and synaptic connections, the
multiple trace theory seems to be a more reasonable one. Of course, these
two theories are not mutually exclusive in that a combination of the two
is also a possibility.
       A few additional bits of information should be introduced before we
proceed with the development of a model of serial memory.
       Each individual has a span of memory; for adults this memory span
is about seven items. This means that we can reliably repeat or remember
about seven things we have just heard or seen; thus it is not difficult
for us to remember a telephone number but, add the area code and it becomes quite difficult. (The span of an individual's memory is only
mildly correlated with intelligence as measured by I.Q. indices.) There
may be no correlation between the memory span and the rate at which an
individual learns. Nevertheless, the memory span is a subject-dependent
variable and must be taken into account in our model of serial learning.

It was suggested by W. James in 1890 that two types of memory determine the memory span: The primary memory (PM) which was the psychological now and the secondary memory (SM) which was the psychological past. The PM consists of events of which we are presently conscious, and SM consists of events which were but are no longer in the consciousness. A diagram of this idea is shown in Figure 5.

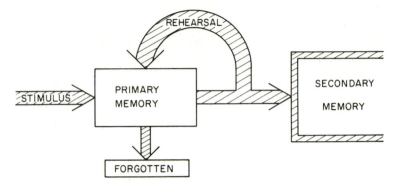

Figure 5

We note in Figure 5 two additions that have not previously been discussed, that of rehearsal and forgetting. These notions intrude since it seems that most items being learned will not get into SM unless they are rehearsed in PM[6] or, to state it even more strongly, it is almost impossible not to rehearse something that one is trying to memorize. The implication is that the PM is a limited storage device and that most items which enter it are quickly forgotten. Further, in order to learn something, i.e., transfer information from PM to SM, one must begin a process of rehearsal. The above sketch, along with assumptions about the independence of rehearsing items in the list and a constant probability of a rehearsed item entering SM, constitute a model of PM used by N. C. Waugh and D. A. Norman.[6] We will not use this model except insofar as Figure 5 is concerned.

In the context of this model (Figure 5), it is easy to visualize the effect of interference. Consider, for example, looking up a phone number. Generally a person will repeat the number to himself, either silently or aloud, while he is going from the phone book to the phone. This repetition of the phone number is the rehearsal in PM. At this point, just as we begin to dial, a sudden noise or other distraction apparently erases the number from our minds. This is an example of external interference. A second example, which most of us have experienced, is adding a long list of numbers or counting a large number of coins. It may be completely quiet and yet, part way through the process, we lose count of what we were doing. This is an example of internal interference.

The relation of the involvement of primary and secondary memory will depend upon the task parameters. The above notion of interference should then have a simple interpretation within the context of our model. Referring again to Figure 2, we see that one would expect the interference to be approximately monotonic. Any accuracy curve, therefore, may or may not depend upon the interaction of two types of memory systems. In the model we made a distinction between items that may be in PM or those in SM. (An item can be in both places at once.)

The qualitative predictions generated from our model are based on set theory. We define the set of interest simply as the "items" which are presented. From this set we define a number of subsets based on the following assumptions:

(A1) The subsets are ordered.
(A2) The subsets have a maximum size k.

The first assumption is based on the method of presentation of the items. Because the items are ordered in time, we assume that subsets of these items maintain this ordering*. The second assumption is based on PM being a limited storage device so that there is an upper limit to the number of items which may be in rehearsal.

For clarity let us consider the set Γ whose elements are from an eight-item list A, B, C, D, E, F, G, and H.

$$\Gamma \equiv \{A\ B\ C\ D\ E\ F\ G\ H\}$$

Because the list is presented one at a time, an individual does not have the set Γ entering PM all at once. Instead it is assumed that the items as they are read form the following subsets:

1. $\gamma_1 \equiv \{A\}$
2. $\gamma_2 \equiv \{AB\}$
3. $\gamma_3 \equiv \{ABC\}$
4. $\gamma_4 \equiv \{ABCD\}$
5. $\gamma_5 \equiv \{ABCDE\}$
6. $\gamma_6 \equiv \{\ BCDEF\}$
7. $\gamma_7 \equiv \{\ \ CDEFG\}$
8. $\gamma_8 \equiv \{\ \ \ \ DEFGH\}$

We note that each of the subsets γ_i are ordered (A1) and that the maximum size of any γ_i is k (A2) where k = 5.** Further, because of the finite capacity of PM, the initial letter (A) is forgotten (removed from PM) in γ_6 as the sixth letter (F) is read. This is not an unreasonable assumption if we assume that the first five γ_i's have previously been rehearsed and, therefore, may be in SM.

To make our model concrete, we are assuming that memory is a consequence of association formation. Further, these associations are formed by set multiplication of the γ_i's in the rehearsal part of PM. Consider the first cycle of rehearsal where the single subset γ_i is present. In cycling the new set, Σ_1 is formed where $\Sigma_1 \equiv \gamma_1 \otimes \gamma_1$.*** The set Σ_1 consists of the element $\{AA\}$, that is, the set Σ_1 is the set of binary associations formed from the subset γ_1. Consider some later time when the subset γ_2 is formed and enters rehearsal. The subsets γ_1 and Σ_1 are already in rehearsal so that the second set of binary associations Σ_2 is formed. This set Σ_2 consists of the products

$$\Sigma_2 = \gamma_2 \otimes \gamma_2 + 2\gamma_1 \otimes \gamma_2$$

_____

*Violations of this assumption occur without question in real behavior and are of genuine interest. They will not be discussed here because of lack of space, however.

**It is found experimentally that k is generally between 5 and 7.[2] ($5 \leq k \leq 7$).

***The symbol (A $\otimes$ B) denotes the direct product of the elements of the sets A and B, with the order maintained.

Note that γ_1 precedes γ_2 in the direct product. The elements of Σ_2 are {AA, AB, BA, BB}.

This process of association formation by set multiplication continues as the list is read. The sets of binary associations Σ_i, $1 < i < 8$, maintains the temporal ordering of the original list. We note that tertiary and higher associations, which would occur from $\Sigma_i \otimes \Sigma_j$ associations ($1 < i$, $j < 8$), are deleted from the above model, based on the notion that these effects are much weaker than the binary effects.

We may now construct a set S from the sets Σ_i such that S contains all the sets Σ_i from $i = 1$ to $i = 8$. The set S can now be used to understand the notion of serial memory. To do this we wish to construct a matrix in which the number of times a particular ordered pair occurring in S will be listed. Such a matrix is given below where the initial column is the first letter of an association and the first row is the second letter of an association. It is clear from this matrix that the symmetric curve that resulted from Hull's analysis of serial memory has been superceded since our matrix is non-symmetrical. For example, there are 34 AB associations, but only 16 BA associations.

	A	B	C	D	E	F	G	H	Sum
A	25	34	41	46	39	30	20	10	245
B	16	25	34	41	36	29	20	10	211
C	9	16	25	34	31	26	19	10	170
D	4	9	16	25	24	21	16	9	124
E	1	4	9	16	16	15	12	7	80
F	0	1	4	9	9	9	8	5	45
G	0	0	1	4	4	4	4	3	20
H	0	0	0	1	1	1	1	1	5
Sum	55	89	130	176	160	135	100	55	900

The elements of the above matrix are obtained by simply counting the number of elements in the subset multiplication. Consider the two subsets γ_j and γ_{j+m} which are represented by two blocks of length p and q, respectively,

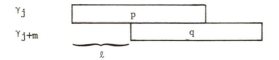

with an overlap interval of length (p-ℓ). The product of these two subsets yields 2pq ordered elements. A little thought shows that the number of pairs with an element of γ_j prior to γ_{j+m} is,

$$\text{number of forward associations} = 2pq - (p-\ell)^2$$

and that the number of pairs with an element of γ_{j+m} prior to an element of γ_j is

$$\text{number of backward associations} = (p-\ell)^2.$$

The sum of the forward and backward association is, of course, 2pq.

The remaining step in the formulation of our model is to interpret this matrix so as to be utilizable in the context of serial memory. To do this we associate the number of times a particular ordered association occurs with the strength of memory of that particular association. For convenience we rewrite the column (C) and row (R) labeled sum in terms of proportions, in the two vectors

$$
C = \begin{bmatrix} .061 \\ .098 \\ .144 \\ .195 \\ .177 \\ .150 \\ .111 \\ .061 \end{bmatrix}
\qquad
R = \begin{bmatrix} .272 \\ .234 \\ .188 \\ .137 \\ .088 \\ .050 \\ .022 \\ .005 \end{bmatrix}
$$

The vectors R and C, therefore, have as elements the strength of forward vs. backward associations, respectively, measured in terms of unitary strength for the entire process. For example, R_1 gives the strength of all binary associations with A as the first element and C_1 gives the strength of all binary associations with A as the second element, and so on down the list.

The problem of facilitation or disruption of associations can now be dealt with simply by suggesting that consistent use of a forward direction is facilitory, but admixture or backward association is disruptive. That is, the problem is related to the task.

To clarify this last remark, let us look at the elements of the vector R plotted against the position of the letters in a ten-item list. This is done in Figure 6 where the rehearsal parameter k assumes the three values k = 5, 6, 7. It is clear that the vector R is essentially

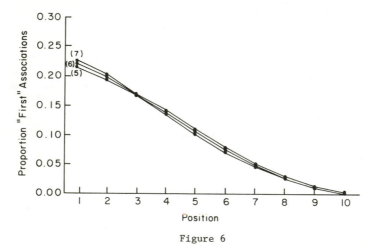

Figure 6

invariant to changes in the parameter k. This would seem to indicate that the relative strengths of forward association does not vary from individual to individual. These functions are similar in shape to those shown in Figure 3.

We now look at a plot of the elements of the vector C versus the position of the letters in a ten-item list in Figure 7. We note that for

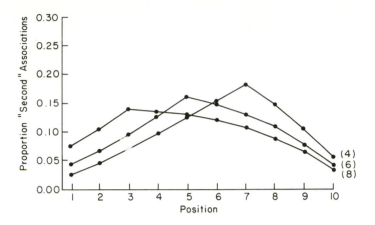

Figure 7

each value of the parameter k, there is a single maximum. The position of this maximum, i.e., at which point in the list it occurs, depends linearly on the value of k. The position of the maximum can be interpreted as the position of maximum interference of the backward associations with the forward associations. This type of interference should disrupt performance, i.e., produce errors. Accuracy is simply the difference between maximum performance and errors; hence, the curves can be inverted to produce accuracy curves. These curves should predict accuracy when background associations are most disruptive as in the anticipation task illustrated in Figure 1.

In a free recall task, subjects do not recall items in a consistent order relative to the stimulus presentation. As one possibility, subjects might use whatever associations are stronger. A theoretical prediction may be obtained if we take the absolute difference between the vectors R and C, i.e., $|R - C|$ and plot this difference as a function of position for two values of the parameter k in Figure 8. This theoretical function is similar in shape to that shown in Figure 2 for free recall (FR).

The real test of any model, aside from its esthetic appeal and qualitative accuracy, is how well it does with new experimental data. Usually the curves published are in terms of averages and, consequently, no information is available about individual subjects. One experiment avoids this by a manipulation of the rate at which the sequence to be learned is presented to the subject. Aaronson[7] trained subjects at one rate and then tested them at a second rate. Using the faster rate should make the list more difficult; moving to a slower rate should make the list easier for all subjects. Figure 9 gives the results of this experiment. In Figure 9a the subjects were trained at the rate of six items per second and 1.5 items per second, and both were tested at the rate of three items per second. It is clear from this figure that those subjects who were trained at a higher rate than that at which they were tested had maximum errors before the middle of the list, and those that were trained at a slower rate than that of testing had a maximum error after the middle of the list. The same phenomenon occurred in Figure 9b where all subjects were trained at the same rate (three items per second), but were tested at two

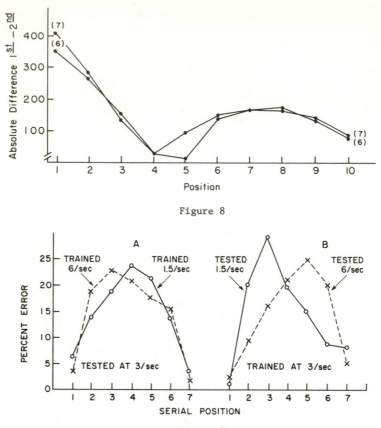

Figure 8

Figure 9

different rates. One's intuitive notion of learning is therefore supported by these results.

In our model we have assumed that the subsets are ordered. It is
entirely possible that items are inverted or misordered. This is more
likely to happen within a subset than between two subsets. Aaronson's
data[7] confirm this, as do some of my own.[2] The pattern of error is not
random. More adjacent inversions are observed and fewer distant inversions are observed than would be expected by chance.

A further correspondence of the model with experimental results occurs if one considers additional free recall experiments. It was determined by Murdock[8] that the proportional accuracy of the first item changes
as a function of list length. This result can be predicted as a direct
function of the proportion of first associations. Figure 6 shows these
for a ten-item list. It can be shown that the proportion of associations
for the first position is higher for an eight-item list than shown, and
lower for a twelve-item list than shown.

In summary, the qualitative predictions of the model are reasonably
successful. This is encouraging and suggests that further empirical work
based on this model might be fruitful in understanding serial memory. For

example, one question would involve exploring reorganizations of subsets in memory (violations of Assumption A1). Empirical evidence will no doubt force modifications and extensions of the model, but this is the most important aspect of a model of this type, to raise questions that might not otherwise be raised.

REFERENCES

1. J. W. McCrary and W. S. Hunter, "Serial Position Curves in Verbal Learning,"*Science* 117 (1953), 131-134.
2. A. O. Dick, "Recognition Serial Learning: Elimination of Item Recall and the Bowed Serial Curve." Unpublished experiment, Lake Forest College, 1968; "The Effect of Type of Recall and Rate of Recall in One-Trial Serial Learning." Unpublished experiment, University of Rochester, 1969.
3. H. Ebbinghaus, "Uber das Geduchtnis,"Leipzeig: Duncher, 1885. (Cited in R. S. Woodworth, and H. Schlosberg, *Experimental Psychology* (New York, 1954).
4. C. L. Hull, "The Conflicting Psychologies of Learning - A Way Out," *Psychological Review* 42 (1935), 730-734.
5. D. O. Hebb, "Distinctive Features of Learning in the Higher Animal." In J. F. Delafresnage, ed., *Brain Mechanisms and Learning* (Oxford).
6. N. C. Waugh and D. A. Norman, "Primary Memory," *Psychological Review* 72 (1965), 89-104.
7. D. Aaronson, "The Temporal Course of Perception in an Immediate Recall Task," *Journal of Experimental Psychology* 76 (1968), 129-140.

A SIMPLE ALGORITHM FOR CONTRACT ACCEPTANCE

Julian Keilson

(Adapted by Bruce J. West)

University of Rochester

This lecture will concern itself with the material presented in reference 1 where an algorithm for contract acceptance is developed. Let us begin by saying that an algorithm is a recipe or set of rules by which one may compute a desired quantity. We will, therefore, concern ourselves with developing a procedure that a contractor may use in making selections between contracts offered him. Of course a problem of this generality does not lend itself readily to mathematical analysis, so one must construct an idealized business environment in which our contractor will function. The environment may or may not correspond to the "real" situation. The present formulation is a first step and is a compromise between fidelity and mathematical tractibility.

In general, the environment for the contractor will consist of periods in which he is idle, interspersed with periods in which he is working on a contract. The acceptance of a job will tie up the contractor for a specific period of time during which he cannot accept another offer. If he accepts a poor job, i.e., one which is low-paying and/or ties him up for an extended period, he might well be engaged when a more profitable job is offered. If, however, he turns down relatively poor jobs, he may go out of business waiting for a good one to come along. This, then is the contractor's dilemma. How can he choose between the jobs offered so as to maximize the average amount of money received per average interval of time worked?

Our first idealization* of the above problem is to assume that all contracts are offered independently. Given this assumption, each opportunity appears to occur without memory of when the preceding opportunities occurred. We consider these events** to occur singly in time and such that the probability of an event occurring in an interval of time Δt is proportional to the time interval. The probability that no contract is offered in the time interval $(t, t+\Delta t)$ is therefore

$$\text{Prob. } \{N(t,t+\Delta t) = 0\} = 1 - \lambda \Delta t + o(\Delta t) \tag{1}$$

---

*The idealization employed is known to probabilists as a Poisson process.

**We will use the terms contract offer, opportunity, and event, interchangeably.

where $N(t, t+\Delta t)$ is the number of events in the interval $(t, t+\Delta t)$ and λ is the rate constant characterizing the process, i.e., the average number of events per unit time. The quantity (Δt) is a term that goes to zero faster than Δt as Δt goes to zero, i.e.,

$$\lim_{\Delta t \to 0} \frac{1}{\Delta t} \, o(\Delta t) \to 0 \; .$$

Let us further define τ to be the random time between two consecutive job opportunities and let the probability distribution for τ be

$$P(t) = \text{Prob.} \; (\tau > t)$$

Then for a positive increment of time Δt, it is clear that

$$P(t+\Delta t) = \text{Prob.} \; (\tau > t+\Delta t)$$

where the right-hand side of this equation may be factored as follows

$$\text{Prob.} \; (\tau > t+\Delta t) = \text{Prob.} \; (\tau > t) \; \text{Prob.} \; [N(t, t+\Delta t) = 0 | \tau > t]. \qquad (2)$$

Thus, the probability that the time (τ) between two consecutive events is greater than $(t+\Delta t)$ is equal to the probability that τ is greater than t, times the probability that no event occurs between t and $t+\Delta t$ given that τ is greater than t.

The second term on the right-hand side of equation (2) is a conditional probability requiring that $\tau > t$. However, since the process under consideration does not "remember" what occurred prior to the interval of interest $(t, t+\Delta t)$, we may replace this conditional probability with the unconditional probability of equation (1). This yields the relation

$$P(t+\Delta t) = P(t)[1 - \lambda \Delta t + o(\Delta t)] \qquad (3a)$$

or, in the limit of very small time intervals,

$$\lim_{\Delta t \to 0} \frac{P(t+\Delta t) - P(t)}{\Delta t} = \frac{d}{dt} P(t) = - \lambda P(t). \qquad (3b)$$

The normalized solution to equation (3b) is

$$P(t) = \exp\{-\lambda t\} \qquad (3c)$$

and the distribution function for the time between two consecutive job involvements (τ) is

$$\text{Prob.} \; (\tau > t) = 1 - e^{-\lambda t} \qquad (4)$$

which depends only on the rate at which opportunities present themselves (λ) and the time interval (t). Arrivals subject to this distribution [equation (4)] are said to occur at Poisson epochs of rate λ.

If the contractor accepts only a single type of contract in the above environment, opportunities of this type will present themselves at a rate λ. The time between two such offers will have the probability distribution given by equation (4). This "exponential" distribution is well known and has been applied in numerous contexts, e.g., see Feller.[2]

We may now generalize the above situation so that the contractor is allowed to accept a number of different jobs. Let there be a number N of job types j which the contractor will accept, $(j = 1, 2, \ldots, N)$. Opportunties of type j present themselves to the contractor at random Poisson epochs of rate λ_j and become unavailable if not immediately accepted. We may distinguish each event by their reward s_j and involvement time τ_j, both of which are considered to be random variables. The expected payoff S_j and average involvement time T_j are given by the expectation values,

$$S_j = E\{s_j\} \tag{5a}$$

$$T_j = E\{\tau_j\} \tag{5b}$$

and are considered independent.

In our carefully constructed business environment, the contractor is restricted to accepting only one contract at a time and not accepting a second until the first is completed. If undertaken, poor contracts tie him up and deny him access to better opportunities. To reiterate our general problem, we must supply the contractor with a procedure for accepting or rejecting opportunities of a given type. A logical criterion* would be one which would permit him to maximize his income rate over a long period of time.

If we specify a policy by stating the set A of job types which our contractor will accept, we shall establish that his average income will be

$$I(A) = \sum_{j \epsilon A} \lambda_j S_j / [1 + \sum_{j \epsilon A} \lambda_j T_j] \tag{6a}$$

where the notation $j \epsilon A$ means that j is a member of A. Our problem is then to find that set of contacts A which maximizes the contractor's average income I(A) as given by equation (6a).

The expression for the average income I(A) may be obtained in the following way. The contractor will either be in an idle state or he will be engaged in a contract from the set A. When he is idle there is a probability per unit time λ_1 that a contract of type j = 1 will present itself, a probability per unit time λ_2 that a contract of type j = 2 will present itself, etc. In total, there is a probability per unit time λ_A where

$$\lambda_A = \sum_{j \epsilon A} \lambda_j \tag{7}$$

of any contract from the set A presenting itself.

Equation (7) becomes more apparent if we consider the probability that no contracts are offered in the time interval (t,t+Δt) in which he is idle. We write this probability in terms of the probabilities for the independent job types in the set A as

$$\text{Prob. } \{N(t,t+\Delta t) = 0\} = \prod_{j \epsilon A} \text{Prob. } \{N^{(j)}(t,t+\Delta t) = 0\} \tag{8a}$$

where N is the total number of jobs offered and $N^{(j)}$ is the number of jobs of type j. By using equation (1), equation (8a) may be rewritten as

$$\text{Prob. } \{N(t,t+\Delta t) = 0\} = \prod_{j \epsilon A} \{1 - \lambda_j \Delta t + o(\Delta t)\} \tag{8b}$$

If we retain only those terms which are of first order in Δt, the finite

---

*It can be shown that this criterion also optimizes the long term discounted future income of the contractor.

product in equation (8b) becomes

$$\text{Prob. } [N(t,t+\Delta t) = 0] = 1 - \sum_{j \varepsilon A} \lambda_j \Delta t + o(\Delta t) . \qquad (8c)$$

We now see that the overall process described by equation (8c) has the same form as the component processes described by equation (1) except that the λ_j associated with each of the latter has been replaced by an overall λ_A as defined by equation (7). One may, therefore, resubmit the arguments which lead from equation (1) to equations (3) and (4) and from these obtain the probability distribution for the total time between contract offers (T) of any type. For the general case this is

$$\text{Prob. } \{T<t\} = 1 - e^{-\lambda_A t} . \qquad (9a)$$

The probability density for the time between events is, therefore,

$$f_A(t) = \lambda_A e^{-\lambda_A t} \qquad (9b)$$

and the average time between contract offers of any type is

$$E[t] = \int_0^\infty t\lambda_A e^{-\lambda_A t} \, dt = \frac{1}{\lambda_A} \int_0^\infty y e^{-y} \, dy = \frac{1}{\lambda_A} . \qquad (9c)$$

Consider a large sequence (M) of intervals, each of which contains an idle period followed by an engagement. In this sequence the number of jobs of type j is M_j with an average involvement time T_j. The total time engaged is, therefore, $\sum_{j \varepsilon A} M_j T_j$. Also, since $1/\lambda_A$ is the average length of time between engagements of any kind and there are M idle periods, the total time spent idle is M/λ_A.* The total duration (D) of the sequence is, therefore,

$$D = M/\lambda_A + \sum_{j \varepsilon A} M_j T_j . \qquad (10)$$

The total expected payoff for the contracts may be obtained in a similar manner,

$$P = \sum_{j \varepsilon A} M_j S_j \qquad (11)$$

which is the average expected payoff for a contract (j) in the set A times the number of contracts of that type which are accepted, summed over the set A. We may write the average income for this sequence as

$$I(A) = P/D$$
$$= \sum_{j \varepsilon A} M_j S_j \left(\frac{M}{\lambda_A} + \sum_{j \varepsilon A} M_j T_j \right)^{-1} . \qquad (12)$$

The above arguments imply that the objective function I(A) will remain valid even if S_j and τ_j are jointly distributed (non-independent). To obtain the form of the objective function given by equation (6), we must

---

*These statements may be made rigorous by qualifying our remarks with the word "asymptotically."

find an expression for the number of jobs of type j in the sequence M, i.e., M_j

We have determined that for a very long sequence of intervals, i.e., M very large, the average time the contractor spends idle is $\overline{\Delta T} = M/\lambda_A$. Now as the average number of jobs of type j offered in a time interval Δt is $\lambda_j \Delta t$, if we write Δt_k as the time interval between the $(k-1)^{st}$ job completion and the kth job offer, then we may write for the total number of jobs of type j,

$$M_j = \lambda_j \sum_{k=1}^{M} \Delta t_k \; . \tag{13a}$$

For a very long sequence the sum (13a) may be reexpressed as

$$\sum_{k=1}^{M} \Delta t_k = M \; \overline{\Delta t} \; .$$

We may therefore write equation (13a) as

$$M_j = \lambda_j M \; \overline{\Delta t} \tag{13b}$$

But we know that the average interval of time between contract offers is $1/\lambda_A$ and, therefore, that

$$M_j = \frac{\lambda_j}{\lambda_A} M \qquad * \tag{13c}$$

which, upon substitution into equation (12), yields

$$I(A) = \frac{\displaystyle\sum_{j \in A} \lambda_j S_j}{1 + \displaystyle\sum_{j \in A} \lambda_j T_j} \tag{6a}$$

as the desired expression for the contractor's income rate over a long period of time.

The problem of finding an optimal set A, the set of jobs which the contractor should accept in order to maximize his average income $I(A)$, may be attacked in the following way. Define the vectors

$$a \equiv (a_1, a_2, \ldots, a_n) \equiv (\lambda_1 S_1, \lambda_2 S_2, \ldots, \lambda_n S_n)$$

and

$$b \equiv (b_1, b_2, \ldots, b_n) \equiv (\lambda_1 T_1, \lambda_2 T_2, \ldots, \lambda_n T_n)$$

where $a_j > 0$ and $b_j > 0$. We will also define a vector μ_A which will be an indicator vector for the set A. Thus we let $\mu_{A_j} = 1$ if j is in the set A($j \in A$) and $\mu_{A_j} = 0$ if j is not in the set A($j \notin A$). Then we may rewrite equation (6) in terms of a, b and μ_A as

$$I(A) = \frac{\mu_A \cdot \alpha}{1 + \mu_A \cdot b} \tag{6b}$$

---

*A more rigorous argument may be constructed along the same lines.

and one must determine the indicator vector μ_A which maximizes $I(A)$.

One may now determine the optimal acceptance set A with the help of the following algorithm. For the sake of simplicity let the set of numbers $\gamma_j = a_j/b_j = S_j/T_j$ be enumerated in descending order so that $\gamma_j > \gamma_k$ when $k > j$. The set γ_j consequently orders the average payoff per unit time worked. Let γ_{N+1} be defined for convenience to have the value zero. Further, let $A_k = \{1,2,\ldots,k\}$ for any positive integer k, and let m be the smallest positive integer for which

$$I(A_m) = \frac{\sum\limits_{1}^{m} a_m}{1 + \sum\limits_{1}^{m} b_m} > \gamma_{m+1} \ . \tag{14}$$

We shall then assert that A_m is an optimal acceptance set. In words, euation (14) asserts that the income rate $I(A_m)$ for the set of the first m job types will be greater than the average payoff per unit time worked for the set of $(m+1)^{st}$-type jobs. Further, if m is the smallest integer for which is this true, then A_m is an optimal set. Note that the optimal set need not be unique. If, for example,

$$I(A_{m-1}) = a_m/b_m, \text{ then } I(A_m) = I(A_{m-1}) \ .$$

The proof of the algorithm [equation (14)] starts with the familiar fact that the inequality $(A/B) > (C/D)$ implies $(A/B) > (A+C)/(B+D)*$ and that strict inequality holds in the second relation whenever the same is true of the first relation. Consequently, for any non-empty subset B of $(m+1, m+2, \ldots, N)$ one will have

$$\frac{a_{m+1}}{b_{m+1}} > \max_B \left\{ \frac{a_j}{b_j} \right\} > \sum_B a_j / \sum_B b_j \tag{15}$$

Equation (15) is a statement of the fact that we have ordered the γ_j such that $\gamma_j > \gamma_{j+1}$. Therefore, γ_{m+1} is greater than or equal to the maximum γ_j in the set B, $(\gamma_m \equiv a_m/b_m)$. The second inequality simply follows from the footnote*, having utilized all the members of the set B. From equations (14) and (15), it then follows that

$$I(A_m) > \sum_B a_j / \sum_B b_j$$

and, hence,

$$I(A_m) = \frac{\sum\limits_{1}^{m} a_j}{1 + \sum\limits_{1}^{m} b_j} > \frac{\sum\limits_{1}^{m} a_j + \sum\limits_{B} a_j}{1 + \sum\limits_{1}^{m} b_j + \sum\limits_{B} b_j} \tag{16}$$

[The income $I(A_m)$ therefore exceeds $I(A)$ for every larger set A containing A_m.] It remains to be shown that the income rate $I(A_m)$ is larger

---

*One can add a constant term to each side of the first inequality $[(A/B) > (C/D)]$ without changing its value. If we add A/D to each side of the inequality and regroup the terms to form $A/B(1+B/D) > (A+C)/D$, the required result follows immediately.

than the income rate $I(C)$ for any subset C of A_m, i.e., that $I(A_m) > I(C)$ for every subset C of A_m

Consider the set C which is a subset of A_m and contains all the contracts $(1,2,\ldots,K-1)$. Thus, C has its first gap, i.e., missing element of A_m, at K. C may also have any number of subsequent gaps. Let

$$U_{K-1} = \sum_1^{K-1} a_j \qquad ; \qquad V_{K-1} = \sum_1^{K-1} b_j$$

$$\alpha = \sum_{C-A_{K-1}} a_j \qquad ; \qquad \beta = \sum_{C-A_{K-1}} b_j \ .$$

Consider the set C' obtained from C by adding the first missing contract type K. Then

$$I(C') - I(C) = \frac{U_{K-1} + \alpha + a_K}{1 + V_{K-1} + \beta + b_K} - \frac{U_{K-1} + \alpha}{1 + V_{K-1} + \beta}$$

$$= \frac{a_K(1 + V_{K-1} + \beta) - b_K(U_{K-1} + \alpha)}{D} \qquad (17)$$

where D is positive. Since $K < m$ and m is the smallest value for which equation (14) is true, we have

$$\frac{a_K}{b_K} > \frac{U_{K-1}}{1 + V_{K-1}}$$

and, following the reasoning of equation (15), we also have

$$\frac{a_K}{b_K} > \frac{\alpha}{\beta} \ .$$

Hence the numerator of equation (17) is non-negative and the income of C' exceeds that of C, i.e., $I(C') > I(C)$. When we fill in the next lowest gap in A_m for C', we obtain $I(C'') > I(C')$ by exactly the same reasoning and finally obtain $I(A_m)$. Hence, $I(A_m) > I(C)$ for every proper subset C of A_m.

It is readily shown* that $I(A_K)$ is monotonic nondecreasing in the range $1 < K < m$; that is, although the individual γ_K's which are incorporated in $I(A_K)$ are becoming smaller, $I(A_K)$ itself is increasing. Correspondingly in the range $m < K < N$, $I(A_K)$ is a monotonic nonincreasing function. In short, the function $I(A_K)$ reaches its maximum value for $K = m$ and, therefore A_m is the optimal set.

Let us now examine some of the interesting consequences of having established this algorithm. These can be stated as the following corollaries:

1. $a_1/(1+b_1)$ is a necessary and sufficient condition that the optimal policy be acceptance of type (1) contracts only. The apparent simplicity of this result should not lead one to think that it is trivial. For if one has an ordered set of contracts

---

*For $\frac{A}{B} > \frac{C}{D} => \frac{A}{B} > \frac{A+C}{B+D} > \frac{C}{D}$. Hence, $\frac{A}{B} > \frac{C}{D} > \frac{E}{F} => \frac{B+D}{B+D} > \frac{A+C+E}{B+D+F}$, and the assertion follows.

$\gamma_1 > \gamma_2 > \ldots > \gamma_N$ and the above relation is satisfied, the the optimal policy criteria imposes specialization rather than diversification as a policy for maximizing income.

One may speculate on other kinds of situations where an individual finds himself in the position of the contractor. For example, a person engaged in research may find that interesting new problems present themselves at random Poisson epochs, and he must decide which of the problems presented should be explored. A suitably defined γ (perhaps the average number of patents per average unit of involvement time) may determine whether it is more profitable, in terms of constructive work accomplished, for him to specialize in a particular field or to branch out into a number of fields.

2. A necessary and sufficient condition that all the contracts be accepted is that

$$I(A_K) = \frac{\sum\limits_{1}^{K} a_j}{1 + \sum\limits_{1}^{K} b_j}$$

be a monotonic nondecreasing function of K over the range $1 < K < N$. It is interesting to note that if one considers a conglomerate company which is buying new businesses, one might choose such a model. For example, if the γ_j are defined as the average return per unit expenditure in purchasing a new company (j), then corollary (2) would be applicable to such a company. In this example, the algorithm would determine the company's maximum size, i.e., how many companies would eventually belong to the conglomerate.

3. $I(A_K)$ is constant for all K, if and only if $\frac{a_1}{1+b_1} = \frac{a_j}{b_j}$ for all $j > 2$. That is, the contractor's income is independent of his decision policy only if $\frac{a_1}{1+b_1} = \frac{a_j}{b_j}$ for all $j > 2$.

Until now we have restricted ourselves to either a complete acceptance or rejection of all opportunities of a given type. One might also wish to consider the case where the contractor only accepts a certain fraction of a given type opportunity (λ_j) from among a long series of such offers, rather than accepting all or none of them. Such a procedure is termed a mixed strategy and is realized by associating with the contract of type j, a probability p_j for acceptance. It will be shown next that such a mixed strategy cannot increase the expected income.

The use of a probability weighted acceptance of a contract of type j replaces λ_j by $p_j\lambda_j$; the indicator vector μ_A in equation (6b) no longer has the values ($\mu_{A_j} = 1$, $j\epsilon A$) and ($\mu_{A_j} = 0$, $j\epsilon A$) but rather $\mu_{A_j} = P_j$, so that equation (6b) becomes

$$I(p) = \frac{p \cdot a}{1 + p \cdot b} \tag{18a}$$

where as previously, $a_j = \lambda_j S_j$ and $b_j = \lambda_j T_j$. The vector p must be chosen subject to the constraints

$$0 < p_j < 1 \qquad , \qquad j = 1,2,\ldots,N . \tag{18b}$$

Suppose that p is such that I(p) is a maximum. For an infinitesimal change δ_p of p, one has from equation (18a)

$$\delta I(p) = \delta p \frac{a}{1 + p \cdot b} - b \frac{p \cdot a}{(1 + p \cdot b)^2} \tag{19a}$$

In particular, when all coordinates p_j are held fixed except p_k, one has

$$(1 + p \cdot b)\delta I(p) = \delta p_k \{ a_k - b_k \frac{p \cdot a}{1 + p \cdot b} = \delta p_k \{ a_k - b_k I(p) \} \tag{19b}$$

By hypothesis, I(p) is a maximum which implies that $\delta I(p)$ may only be negative. Then if for some k, $\{a_k - b_k I(p)\}$ is positive, p_k must have the value $p_k = 1$. This violates the constraint given in equation (18a) and, therefore, implies that δp_k must be negative. Similarly, if $\{a_k - b_k I(p)\}$ is negative, one must have $p_k = 0$ which, in turn, rules out negative values of δp_k. We then infer that one must have $p_k = 0$, $p_k = 1$, or

$$\frac{p \cdot a}{1 + p \cdot b} = \frac{a_k}{b_k} \tag{20}$$

for any component k for which p_k does not have the value zero or one. Equation (20) expresses the fact that $\delta I(\vec{p}) = 0$ in equation (19b) if p_k does not have the values zero or unity. One may therefore rewrite equation (18a) as

$$I(\vec{p}) = \frac{p \cdot a}{1 + p \cdot b} = \frac{p \cdot a \sum_k (1-p_k)a_k}{1 + p \cdot b + \sum_k (1-p_k)b_k} \tag{21}$$

which simply reduces to equation (6a). We have thus shown that mixed strategies may be ignored in a search for the maximum of I(p).

ACKNOWLEDGEMENTS

The authors would like to thank Wade W. Badger for his critical reading of the manuscript and for his useful comments.

REFERENCES

1. J. Keilson, *A Simple Algorithm for Contract Acceptance*.
2. W. Feller, *An Introduction to Probability Theory and Its Applications*, Vols. 1 and 2 (New York: John Wiley, 1966).
3. S. M. Ross, *Applied Probability Models with Optimization Applications* (Holden-Day, 1970).

NATURAL FORCES AND EXTREME EVENTS

Aaron Budgor and Bruce J. West

University of Rochester

INTRODUCTION

The topic of the seminar that I will present today concerns itself
with a study of some natural phenomena that are considered extreme events.
An extreme event may be thought of as the occurrence of an incident which
deviates greatly from the norm. The essence of this notion is that there
is some phenomenon which exhibits itself in a normal region of fluctuation
as measured by some appropriately chosen variable. On occasion, this
phenomenon may occur far outside its normal region (an extreme event)--
wind gust loads on airplanes in flight, highest temperatures or lowest
pressures in meteorology, floods and droughts in hydrology, and human life
spans all fall in this category--and to be prepared for such an occurrence,
knowledge of the behavior of such events must be understood.
    In 1935 Emil Gumbel, then working in France, derived an expression
for a distribution function

$$\Phi(x) = \exp\{-\exp[-\beta(x-\mu)]\} \tag{1}$$

which he believed would be useful in predicting oldest ages in human life
spans.[1] Subsequently, the Gumbel distribution was applied to other extreme events, examples of which are cited above, and quite reasonable return times were found.* The derivation of the distribution in equation
(1) was based on two underlying assumptions. The first was that, given a
random variable X such as the height of a river or the magnitude of an
earthquake, the initial distribution and its parameters should be station-
ary, i.e., independent of a shift in time. The second assumption was one
of independence on each of the n observations of our random variable X.
This last assumption may or may not be valid. Consider the daily fluctuations in the level of a river. One would find it difficult to maintain
that the day-to-day levels are independent. On the other hand, the magnitude of flooding or snowfalls on a year-to-year basis would seem to
satisfy such an assumption.[2]
    The objective of this talek is to determine whether the Gumbel dis-
tribution really has predictive value for the study of extreme events and
natural phenomena. If not, an alternative picture may be formulated. The
authors have acquired data for the maximum and minimum river heights for
the Nile River Valley over a span of 1300 years.[3] To this date, the best

---

    *A return time is the length of time one would wait between two oc-
currences of an extreme event of a specified magnitude.

predictive model on the height of Nile floods is a mathematical relation
correlating the height of the Nile with the temperatures of Dutch Harbor,
Alaska and Samoa in the Pacific, and the barometric pressure at Port
Darwin, Australia. The predictive lifetime of this relation was not ex-
ceptionally long.

I would now like to relate to you a little about the history and
hydrology of the Nile Valley,[4] which is both interesting in itself and
will be useful in the later analysis.

The Nile River is probably the longest river in the world, measuring 4157 miles from its most distant source to the entry of the Rosetta
Branch in the Mediterranean. Three principal streams form the Nile. The
largest contributor in volume of water supplied is the Blue Nile. Four-
sevenths of the total water supply of the main stream and nearly 70% of
the Nile flood is derived from it. Its reputed source is a spring to the
south of Lake Tana in Ethiopia, but most of its waters are from tributar-
ies which gush down from the Ethiopian plateau. The rainfall there is
especially heavy. On the average, 1200 mm. fall annually. Since it
rarely snows in the Ethiopian mountains, snowfall may be discounted as a
major source of water to the Blue Nile.

The second stream, the White Nile, supplies two-sevenths of the
total water supply of the main stream. It combines with the Sobat in the
Sudan and then with the Blue Nile at Khartoum. The last principal stream,
the Atbara, drains the northwestern part of Ethiopia and joins the main
stream 200 miles north of Khartoum. It contributes the remaining one-
seventh of the total water volume. The Blue Nile and the Atbara bring
most of the soil which makes Egypt cultivatable.

The flooding of the Nile was of the utmost importance to the
Egyptian people. The main Nile from Khartoum to Aswan flowed between
desert terrain with only a narrow strip of vegetation on each bank. From
Aswan to Cairo, the river was bordered by a flood plain of alluvium which
gradually increased to a maximum width of twelve miles. Thus, long before
Herodotus traveled through Egypt in the fifth century B.C., man prayed
for 16 ells. Pliny explained: "12 ells mean hunger, 13 sufficiency, 14
joy, 15 security, and 16 abundance." A system of twenty Nilometers were
erected along different points between Aswan and Cairo to measure the
river level. They were cylindrical marble columns placed in the water by
the river banks, approximately 5.3 meters in height, 1.9 m. in circumference at the base, and 1.6 m. in circumference at the summit. On the north,
south, southeast and east faces, a series of gradations were etched, each
series being calibrated from a different zero-point. A fine degree of
measuring accuracy was therefore obtainable. Two eagles, one of each sex,
were placed at the sumit of a Nilometer and on the first day of the flood,
the Pharoah, his priests, and the populace, deep in prayer, would listen
for which of the two eagles screamed first: if it was the male, a great
flood was predicted and the Pharoah immediately raised the price of the
yet unsown corn. This is probably one of the first documented instances
of religious exploitation of the people.

Using 1300 consecutive years of reliable Nile River measurement
data, it becomes a simple matter to test the quality of the fit of equation (1). At the outset, however, it will be informative to derive the
Gumbel distribution from the process underlying the time-dependent change
of extreme river heights. Thus, one should be able to determine how large
a flood (or drought) one may expect, based on the magnitude of past floods
(or droughts).

THE NATURAL FORCE

Consider the rate equation

$$\frac{dx}{dt} = G(x) + F(t) \tag{2}$$

for an unspecified process as measured by the variable x. The functions $G(x)$ and $F(t)$ are, respectively, a forcing term which determines the rate of growth of the variable x and a random forcing function which disperses the values of x about those given by the deterministic forcing function $G(x)$. We make several standard assumptions about the characteristics of the random forcing term $F(t)$. We shall assume firstly that the average value of $F(t)$ is zero, i.e., $<F(t)> = 0$. Secondly, the self-correlation of $F(t)$ at two different times (t_1, t_2) is proportional to a Dirac delta function in time,

$$<F(t_1)F(t_2)> = \sigma^2 \delta(t_1 - t_2) \tag{3}$$

Equation (3) expresses the fact that the random forcing term has a very short memory. A form of $F(t)$ which satisfies these two conditions is the Gaussian random process which is completely characterized by the parameter σ. These two assumptions are generally made in the Theory of Brownian Motion.[5]

Due to the random nature of the force $F(t)$, we shall examine the probability that our variable has a value x at time t. The probability density $[\phi(x,t)]$ for such a process as defined by equation (2), satisfies the partial differential equation known as the forward Kolmogorov equation,[6]

$$\frac{\partial \phi(x,t)}{\partial t} = -\frac{\partial}{\partial x}\{a(x)\phi(x,t)\} + \frac{1}{2}\frac{\partial^2}{\partial x^2}\{b(x)\phi(x,t)\} \tag{4}$$

where $a(x)$ is the rate of growth of the mean and $b(x)$ is the rate of growth of the variance for the process defined by equation (2). We have from equation (2) that

$$a(x) = \lim_{\Delta t \to 0}\left\{\frac{1}{\Delta t}<\Delta x>\right\} = G(x)$$

$$b(x) = \lim_{\Delta t \to 0}\left\{\frac{1}{\Delta t}<(\Delta x)^2>\right\} = \sigma^2 .$$

When $a(x)$ and $b(x)$ are defined by equations (2) and (3), equation (4) reduces to the Fokker-Planck equation,

$$\frac{\partial \phi(x,t)}{\partial t} = -\frac{\partial}{\partial x}\{G(x)\phi(x,t)\} + \frac{\sigma^2}{2}\frac{\partial^2}{\partial x^2}\phi(x,t) \tag{5}$$

The equilibrium, or time-independent solution of equation (5) may be obtained by setting the time derivative equal to zero, i.e., $\partial\phi/\partial t = 0$ and solving the resulting equation,

$$-\frac{\partial}{\partial x}\{G(x)\phi(x)\} + \frac{\sigma^2}{2}\frac{\partial^2}{\partial x^2}\phi(x) = 0 . \tag{6}$$

The solution to equation (6) is clearly

$$\phi(x) = N \exp \left\{ \frac{2}{\sigma^2} \int_0^x G(\xi)\, d\xi \right\}$$ (7)

where N is the normalization chosen such that

$$\int_0^\infty \phi(x)\, dx = 1$$ (8)

Our procedure is now clear. By selecting

$$G(x) = k[e^{-\beta(x-\mu)} - 1]$$ (9)

where

$$k = 2\sigma^2/\beta$$

and

$$N = \beta\, e^{-e^{-\beta\mu}}\, e^{-\mu}$$

one obtains a probability density for the Gumbel distribution as given by equation (1). Equation (7) then takes the form

$$\phi(x) = \beta\, e^{-\beta(x-\mu)}\, e^{-e^{-\beta(x-\mu)}}.$$ (10)

We may now explore the relationship between the functional form of the driving force [G(x)] in equation (9) and the probability density equation (10).

We noted in the rate equation [equation (2)] that G(x) was a force. It may, therefore, be used to define a potential V(x) such that

$$G(x) = - \frac{\partial V(x)}{\partial x}$$ (11a)

where

$$V(x) = k(x-\mu) + k/\beta\, e^{-\beta(x-\mu)}$$ (11b)

By plotting a graph of both $\phi(x)$ and V(x) vs. x, (cf. Figure 1), one notes that

1) The potential V(x) has a single minimum located at $x = \mu$. This implies that the driving force at $x = \mu$ is zero, i.e., $G(X=\mu) = 0$; and

2) the probability density has a maximum concentration at the potential minimum μ, i.e., the most probable extreme value.

By expanding the driving force in tne neighborhood of the potential minimum ($x = \mu$), we obtain

$$G(x) = k[e^{-\beta(x-\mu)} - 1] \cong - k\beta(x-\mu) + \frac{k\beta^2}{2!} (x-\mu)^2 - \dots .$$

Thus, it is apparent that to first order there is a restoring force $[-k\beta(x-\mu)]$ back to the most probable value of x, i.e., $x = \mu$.

Further insight may be gained as to the relation between the probability density, the natural force, and the potential by considering the average (mean) value of x and the dispersion σ_x of the distribution.

$$<x> = \int_0^\infty x\phi(x)\, dx = \mu + \gamma/\beta$$ (12a)

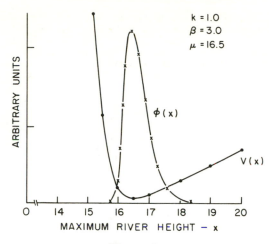

Figure 1

$$\sigma_x^2 = <x^2 - <x>^2> = \pi^2/6\beta^2 \qquad\qquad (12b)$$

where γ is Euler's constant ($\gamma = 0.5772157\ldots\ldots$). From equations (11) and (12a) we see that as $\beta \to 0$ the potential $V(x)$ flattens out, the driving force $G(x)$ goes to zero, and less and less of the probability density $\phi(x)$ is concentrated at the point $x = \mu$. In addition, the average value of $x(<x>)$ and σ_x go to infinity. Conversely, as $\beta \to \infty$ the potential becomes deeper, the driving force approaches a constant $(-k)$, and the probability density peaks more and more at $x = \mu$, i.e., $<x>$ approaches μ and σ_x tends to zero. The characteristics of the natural force (and potential) are therefore completely reflected in the probability density $\phi(x)$.

With the same picture in mind, we now turn to our data in order to determine how well this natural force replicates the geological and meteorological effects which lead to both flood and drought in the Nile River Valley.

RESULTS

The data we use in our analysis give us both the maximum and minimum height of the Nile as measured by the Nilometer at Rodah (Cairo). The data from 621 to 722 A.D. may be suspect, however, because during that time the main measuring station was changed from Memphis, but due to the proximity of the two cities and the similarity of the terrain one may consider these measurements as consistent. The data from the period 1721-1922 are probably less reliable since the method of irrigation was changed. For thousands of years the land was watered by short canals which received water only when the river was in flood. The canals delivered the muddy water to the land which was divided into compartments, or basins, by cross banks running from the river bank to the higher desert edge. With this method the amount of land which could be irrigated was variable and, in times of low flood, resulted in famine. In the middle of the 18th century, a new method of irrigation was developed where every two or three weeks a small quantity of water was run onto the land. Barrages or low dams were built across the Nile to enable its level to be raised sufficiently for continuous flow into the main canals. Branch canals extended from the main canals and these in turn were subdivided

into smaller canals called distributaries which delivered the water to
the irrigation ditches and then onto the land. By the 20th century,
nearly all the irrigation in the Nile Valley was done in this fashion.

To indicate the form in which the data were used, we list in Table
1 the number of times, n_i, in 1300 years that the minimum height of the
river reached a value x_i. The cumulative frequency of the minima between

TABLE 1

x_i	n_i	$\Sigma_i n_i$	$\Sigma_i n_i/N$	x_i	n_i	$\Sigma_i n_i$	$\Sigma_i n_i/N$
9.0	2	2	.0019	12.2	22	765	.7449
9.1	0	2	.0019	12.3	27	792	.7712
9.2	0	2	.0019	12.4	13	805	.7838
9.3	1	3	.0029	12.5	37	842	.8199
9.4	2	5	.0049	12.6	28	870	.8471
9.5	1	6	.0058	12.7	17	887	.8637
9.6	4	10	.0097	12.8	13	900	.8763
9.7	8	18	.0175	12.9	22	922	.8978
9.8	7	25	.0243	13.0	14	936	.9114
9.9	16	41	.0399	13.1	15	951	.9260
10.0	11	52	.0506	13.2	12	963	.9377
10.1	5	57	.0555	13.3	6	969	.9435
10.2	7	64	.0623	13.4	6	975	.9494
10.3	17	81	.0788	13.5	8	983	.9571
10.4	35	116	.1129	13.6	3	986	.9601
10.5	13	129	.1256	13.7	1	987	.9611
10.6	10	139	.1353	13.8	2	989	.9630
10.7	27	166	.1616	13.9	6	995	.9688
10.8	27	193	.1879	14.0	1	996	.9698
10.9	25	218	.2123	14.1	3	999	.9727
11.0	66	284	.2765	14.2	3	1002	.9757
11.1	40	324	.3155	14.3	1	1003	.9766
11.2	33	357	.3476	14.4	1	1004	.9776
11.3	42	399	.3885	14.5	6	1010	.9834
11.4	41	440	.4284	14.6	2	1012	.9854
11.5	68	508	.4946	14.7	4	1016	.9893
11.6	35	543	.5287	14.8	2	1018	.9912
11.7	39	582	.5667	14.9	3	1021	.9941
11.8	42	624	.6076	15.0	2	1023	.9961
11.9	28	652	.6349	15.1	0	1023	.9961
12.0	35	687	.6689	15.2	2	1025	.9981
12.1	56	743	.7235	15.3	1	1026	.9990

the heights x_i and x_{i+1} is also listed and, finally, the cumulative relative frequency is given for a total of N measurements. A plot of the
cumulative frequency is given in Figure 2 where the ordinate is the height
of the river and the abscissa is the Gumbel distribution $\Phi(x)$ as given by
equation (1). On this graph paper, then, a straight line represents a
Gumbel distribution and the slope of the line gives the parameter β in
equation (1). We can see from Figure 2 that the minimum river height does
indeed follow the appropriate distribution.

In Figure 3 we plot the flood heights of the river for the 1300
years. As seen from this figure, the data from flooding fluctuate slowly
about the theoretical straight line $\Phi(x)$, in contrast to the data from
the minimum river height which is much more stable. This would seem to
indicate that our natural force, as given by equation (9), is more appropriate for droughts than for floods.

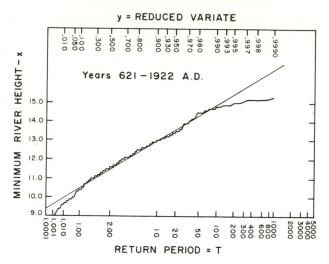

Figure 2

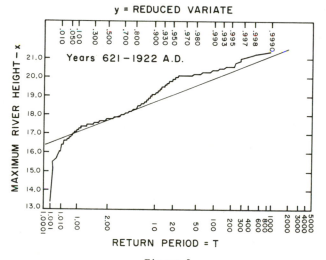

Figure 3

One may use Figures 2 and 3 to predict the maximum and minimum river heights as follows. Consider a minimum river height of 12 meters. By examining the scale on Figure 2 labeled Return Period, we note that, on the average, every 3½ years the river minimum is 12 meters. Correspondingly, the river maximum reaches a value of 20 meters every 200 years. Of course, due to fluctuations in the river maxima, this latter prediction is less reliable than the former.

DISCUSSION

To determine the cause of the fluctuation of the data for the flood heights about the theoretical straight line, let us consider the data century by century. This should reveal any detailed structure which is not evident in Figure 3. A typical plot of the cumulative distribution for a century of data is shown in Figure 4, where we have used the

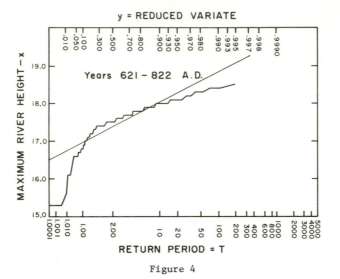

Figure 4

data for the flood heights from 721 to 922 A.D. We can see from this figure that the cumulative distribution for the data points is not at all similar to the Gumbel distribution. The probability accumulates too rapidly for low flood heights, as well as for large flood heights, and resembles the Gumbel distribution in only a small region of intermediate flood values. Furthermore, if we look at the distribution density, i.e., the first derivative of the distribution function, as shown in Figure 5 for the century 1021-1122 A.D., one can see that this density, as well as all the densities common to the first ten centuries of data, are bi-modal. The density function for the river minima remains a Gumbel century by century.

The existence of this secondary maximum in Figure 5 indicates that there is a clustering of events at a flood height lower than would be expected by the Gumbel density. This failure to cut off the lower flood heights indicates that our potential function [equation (11b)] is not applicable in this case, i.e., the potential does not reproduce the system considered. A little thought makes it quite clear that a bi-modal distribution of the form shown in Figure 5 could only arise, in our model, from a potential function with two minima. It is now necessary to return to the discussion of the physical structure of the Nile and determine the appropriateness of this double minima potential.

It is well established, if not well known, that the Blue and the White Niles contribute 70% and 20% of the waters of the main Nile, respectively, with the remaining 10% being contributed by the River Atbara. As the Blue Nile rises in flood each year, it pushes on the waters of the White Nile, forcing them to back up. The White Nile, therefore, becomes

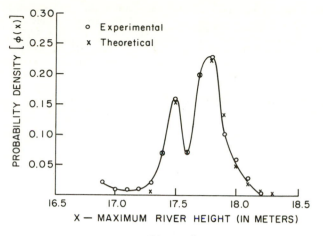

Figure 5

a natural reservoir until the Blue Nile begins to fall in height, at which time the White Nile releases its water. This ponding of the White Nile could have a number of effects on the flood heights measured along the main Nile.

If one were looking at the maximum flood heights of the Blue Nile alone, one would expect to find a smooth distribution of the extreme values, i.e., the Gumbel distribution. The effect on these flood heights produced by the ponding of the White Nile could be any of the following:
1. If a constant quantity of water was backed up to pond the White Nile, independent of the flood height of the Blue Nile, then the distribution of flood heights as measured along the Main Nile would be shifted downward with respect to that of the Blue Nile, but without a substantial change in shape.
2. If the quantity of water backed up to pond the White Nile increased with the flood height of the Blue Nile, then the distribution of flood heights as measured along the main Nile would be skewed, with respect to that of the Blue Nile, toward lower values.

Neither of the above effects would lead to the bi-modal character of the distribution which has been observed. A third alternative, however, does present itself. This is a combination of the above two. Firstly, the quantity of water which is ponded in the White Nile must be an increasing function of the height of the Blue Nile flood. Secondly, this increase must have an upper limit (X=L) after which the White Nile is saturated and a constant amount of water remains independent of the flood height of the Blue Nile (X>L).

We may model this situation by writing the rate equation [equation (2)] as

$$\frac{dx}{dt} = G_1(x) + \{G_2(x) - G_1(x)\}\varepsilon(X-L) + F(t) \tag{13}$$

where $\varepsilon(X-L)$ is the unit step function

$$\varepsilon(X-L) = \begin{cases} 0 & X<L \\ 1 & X>L \end{cases}$$

and where $G_1(x)$ and $G_2(x)$ are two functions which satisfy the conditions established for equation (2). We can now use the Fokker-Planck equation to determine the probability density $[\Phi(x)]$ for the process defined by equation (13). The solution is just that given by equation (7), where now

$$G(\xi) = G_1(\xi) + \{G_2(\xi) - G_1(\xi)\}\varepsilon(\xi-L).$$

For the case of a sharp boundary between regions, the solution to the problem would be,

$$\phi_1(x) = N \exp\{2/\sigma^2 \int_0^X G_1(\xi) \, d\xi\} \quad ; \quad X<L \tag{14a}$$

and

$$\phi_2(x) = N\{\exp 2/\sigma^2 \int_L^X G_2(\xi) \, d\xi\} \quad ; \quad X\geq L \tag{14b}$$

and therefore

$$\phi(x) = \{1 - \varepsilon(X-L)\}\phi_1(x) + \varepsilon(X-L)\phi_2(x) \tag{15}$$

It is clear, however, that no such sharp boundary exists in the real situation. But by using equation (15) as motivation, we may write a solution of the form

$$\phi(x) = c_1\phi_1(x) + c_2\phi_2(x) \tag{16}$$

as a first order approximation to a solution when the boundary becomes a region and not a fixed point. We introduce the notion of a region because the physical structure of the river is not constant in time. The shifting of sand bars, the depositing of silt, etc., all tend to change the value of L from year to year. The functional form of equation (16) differs from that of equation (15) in that the probability densities $\phi_1(x)$ and $\phi_2(x)$ both tend to zero in the neighborhood of L, but may both be non-zero simultaneously, i.e., they may both contribute in this region.

Although equation (16) is not obtained in a mathematically rigorous manner, the authors believe that there is sufficient justification for the use of this form for the probability density. Since there is an overlap region in which both $\phi_1(x)$ and $\phi_2(x)$ contribute, the constants c_1 and c_2 are obtained by fitting the data and by employing the overall normalization condition

$$\int_0^\infty \{c_1\phi_1(x) + c_2\phi_2(x)\} \, dx = 1 \tag{17}$$

Due to the fact that both ϕ_1 and ϕ_2 are descriptive of extreme events, it would not be unlikely that they have a common functional form, but with a different parameterization. We will make the assumption that

$$\phi_1(x) = \beta_1 e^{-\beta_1(x-\mu_1)} \, e^{-e^{-\beta_1(x-\mu_1)}} \tag{18a}$$

and

$$\phi_2(x) = \beta_2 e^{-\beta_2(x-\mu_2)} \, e^{-e^{-\beta_2(x-\mu_2)}} \tag{18b}$$

that is, they both are Gumbel densities. The parameters β_1, μ_1, β_2, and μ_2 can then be obtained by fitting the data.

Further consideration also leads us to expect that $c_2 > c_1$. This occurs because of the contribution of the waters of the River Atbara. Although the River Atbara contributes only about 10% of the water of the main Nile, it makes its contribution over a three to four month period and is usually in phase with the Blue Nile. The common source of the River Atbara and Blue Nile prescribe that when there is a sudden rise in height along the Blue Nile, there is also a corresponding rise in height along the River Atbara. The rise is dampened along the Blue Nile, but not on the River Atbara. Generally the crest of both disturbances meet at the main Nile to enhance the flood height. This superposition of flood heights tends to make large floods more probable and, therefore, $c_2 > c_1$ in equation (16). In fact, by doing a least squares fit of equation (16) where the ϕ_1, ϕ_2 are defined, respectively, by equations (18a,b), it can be shown that for Figure 5, $c_1 = .49$ and $c_2 = .65$.

In contrast to the distribution of maxima, the distribution of minima follow a Gumbel very closely. This would seem to imply that either there is only one distribution function describing river levels, or that the distribution function for the event is a composite of several distributions whose corresponding coefficients, save one, are small.

Intuitively, this seems to be the more likely explanation since it applies equally well for the maxima problem. By examining the fine structure of each process, one should be able to resolve the spectrum of coefficients for every distribution.

ACKNOWLEDGEMENTS

We would like to thank Professor E. W. Montroll for suggesting this problem to us, introducing us to the references, and for many fruitful discussions.

REFERENCES

1. E. Gumbel, *Statistical Theory of Extreme Values and Some Practical Applications*, Applied Mathematics Series 33, National Bureau of Standards (1954).
2. W. W. Badger and E. W. Montroll, *Quantitative Aspects of Social Phenomena*, (Gordon and Breach, New York, 1974).
3. Prince Omar Toussoun, *Memoires sur l'histoire du Nil*, V. 8, 9, 10, Memoires de l'Inst. d'Egypte (1925).
4. H. E. Hurst, *The Nile*, (Constable Press, 1952).
5. M. C. Wang and G. E. Uhlenbeck, *Rev. of Mod. Phys.* 17, 323 (1945).
6. W. Feller, *An Introduction to Probability Theory and Its Applications*, Vol. 2 (John Wiley and Sons, Inc., New York, 1966), p. 326.

SYSTEMS OF MATING

J. H. B. Kemperman

University of Rochester

The general philosophy of this talk will be to consider certain
problems (systems of mating) and to formulate them in a mathematically
precise manner. The assumption entering into the construction of the
model of the problem will be made explicit and solutions will be found
in steps. At each step certain of the initial assumptions will be relaxed so as to make the model more complex and a more reasonable approximation to natural systems of mating. The deductions made within the
model, as well as the assumptions upon which it depends, will be interpreted as much as possible in terms of the natural system under consideration. It will be seen that even in cases where the model system cannot
be solved exactly, or where the solutions do not correspond exactly to
the natural system, it is possible to obtain a great deal of information
about the system.
    The general problem we are interested in is how stable population
patterns are formed in large populations under given systems of mating.
Populations consist of groups of individuals which have common characteristics such as hair or eye coloring, or races in humans, or breeds in
dogs, etc. These groupings in the population will be called types so
that, by a population pattern, we will mean the distribution of these individuals over the different types. A stable population pattern is one
in which the relative number of individuals of each type remains constant
from the parental to filial generation. We may also refer to this pattern
as the equilibrium distribution of types of individuals. The system of
mating referred to above determines the way in which members of the population form parental pairs.
    We will review, by way of introduction, how a certain population
consisting of three types of individuals reaches equilibrium, i.e., a
stable population pattern, using a random system of mating. From this
restricted case we will go on to consider a more general system of mating
for a population of n types of individuals and show what equilibrium situations may be reached. We will also consider certain side conditions
which may make the system of mating more "natural."
    Consider a population of N males and N females (N large) where the
individuals are classified by two genes, A and a, which are not sex linked.
These determine three types of individuals, as shown in Table 1. Since

TABLE 1

Genotype	AA	Aa	aa
Number in each sex:	1	2	3
	N_1	N_2	N_3

47

the genes are not sex linked, we may assume that the type distribution for males coincides with the type distribution for females. Here,

$$\sum_{j=1}^{3} N_j = N \tag{1}$$

The relative number of individuals of type j in the group is given by

$$P_j = N_j/N \tag{2}$$

and, therefore,

$$\sum_{j=1}^{3} P_j = 1 \tag{3}$$

The P_j's indicate the way in which the genotypes considered distribute themselves within the population. In Figure 1 we graphically illustrate the set of all possible distributions (P_1, P_2, P_3) by the shaded area in the (P_1, P_3)-plane, $(P_2 = 1 - P_1 - P_3)$.

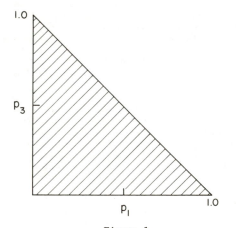

Figure 1

To determine the equilibrium distribution, we note that in any type of mating (random or non-random), the gene ratios are conserved from generation to generation (at least when all individuals get mated). For an arbitrary distribution (P_1, P_2, P_3), the ratios of genes of Type A and a are given by

$$P_A = \frac{2N_1 + N_2}{2N} = P_1 + \tfrac{1}{2}P_2 \tag{4a}$$

$$P_a = \frac{2N_3 + N_2}{2N} = P_3 + \tfrac{1}{2}P_2 \tag{4b}$$

for both the male and female members of the population. Under a system of so-called random mating, the offspring will acquire one gene (of type A or a with frequencies P_A, P_a) from its male parent and, independently, one gene from its female parent, (also of type A or a with frequency P_A or P_a, respectively).

The offspring will, therefore, receive an A gene from both parents with a relative frequency $P_A P_A$, or

$$P_1' = P_A{}^2 \tag{5a}$$

where the prime denotes the filial generation. Similarly, they will receive an a gene from both parents with a relative frequency $P_a P_a$, or

$$P_3' = P_a{}^2 \tag{5b}$$

Further, they will obtain an A gene from the males and an a gene from the females, and vice versa, with a relative frequency

$$P_2' = 2P_A P_a \tag{5c}$$

Note that $P_1' + P_2' + P_3' = (P_A + P_a)^2 = 1$, while $\sqrt{P_1'} + \sqrt{P_3'} = P_A + P_a = 1$.

Therefore, after mating at random, the possible type distributions are restricted to the points of a single curve, the so-called Hardy-Weinberg curve, having the equation

$$\sqrt{P_1} + \sqrt{P_3} = 1 \ . \tag{6}$$

For each gene ratio P_A there is one and only corresponding point on this curve. The distributions with a given gene ratio P_A correspond to the points of the straight line with equation

$$P_1 - P_3 = P_A - P_a = \text{constant},$$

which has a 45° slope. The point (P_1, P_3) must remain on this line from generation to generation in any type of mating (in which all individuals get mated). After a random mating, we have that already the first generation offspring has a Hardy-Weinberg distribution (6) and this distribution will remain unchanged throughout all the subsequent generations (again under random mating).

For instance, suppose we start out with an arbitrary initial distribution P_1, P_2, P_3 having the values .2, .2, and .6. We may then determine the values of P_A and P_a from equation (4) to be .3 and .7. The equilibrium values P_1', P_2', P_3' will then be .09, .42 and .49, or just those given by equation (5). That is, under random mating, the distribution P_1, P_2 and P_3, represented by the point x in Figure 2 will move along the $P_1 - P_3 = $ constant curve to the distribution P_1', P_2', P_3' represented by the point x' on the Hardy-Weinberg curve and there it will remain in all subsequent generations.

One may now go on to ask similar questions for a more complex population which has n types of members. We shall assume a natural division of this population of 2N members into N males and N females (N large). The number of males (females) of type i will be denoted by

$$N_i = NP_i \tag{7a}$$

Further, let

$$N_{ij} = NP_{ij} \tag{7b}$$

be the number of pairs of the type (i,j), (that is, a couple where the male is of type i and the female of type j). For the moment, we do not

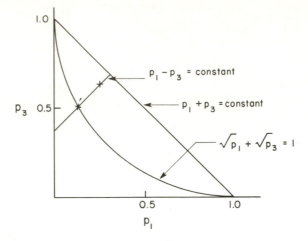

Figure 2

allow any bachelors or spinsters, i.e., no unmated males or females. Then

$$\sum_{j} N_{ij} = N_i \qquad\qquad \text{thus} \qquad\qquad \sum_{j} P_{ij} = P_i , \qquad\qquad (8a)$$

$$\sum_{i} N_{ij} = N_j \qquad\qquad \text{thus} \qquad\qquad \sum_{i} P_{ij} = P_j . \qquad\qquad (8b)$$

Here, the first equation states, for instance, that the number (N_i) of males of type i may be determined by counting all the male-female pairs which involve a male of type i. Given the numbers $N_i = NP_i$, we have that *a priori* any set of non-negative integers $N_{ij} = NP_{ij}$ is possible, provided that the relations (8) are satisfied. Given such a set of numbers N_{ij}, the expected number of offspring of type k is given by

$$N_k' = \sum_{i,j} N_{ij} f_{ijk} \qquad\qquad (9a)$$

where f_{ijk} is the expected proportion of children which will be of type k when the parents are of type (i,j). The type distribution of the children would be

$$P_k' = \sum_{i,j} P_{ij} f_{ijk}. \qquad\qquad (9b)$$

Here we are thinking of very large populations so that it is allowed to restrict ourselves to the deterministic problem which uses only expectations, or mean values, for the relative frequencies of a given type of individual. Further, instead of renormalizing the relative frequencies by the new population size (N') in each generation, we stabilize the population size by assuming for convenience that each parent pair produces exactly one boy and exactly one girl. If the parents are of type (i,j), then the boy will be of type k with probability f_{ijk}; similarly for the girl. In particular,

$$\sum_k f_{ijk} = 1 \tag{10}$$

must hold for each pair (i,j).

As is clear from (9), the type distribution $P' = (P_1,...,P_n)$ of the offspring cannot be computed from the parent-type distribution $P = (P_1,...,P_n)$ unless one also knows how the $N_i = NP_i$ males of type i and the $N_j = NP_j$ females of type j $(i = 1,...,n; j = 1,...,n)$ pair up. In other words, one must know what to choose for the $N_{ij} = NP_{ij}$. Let us define a system of mating as a function G which assigns to every type distribution $P = (P_1,...,P_n)$ a set of non-negative numbers $P_{ij} = G_{ij}(P)$ satisfying (8). Necessarily,

$$\sum_j G_{ij}(P) = P_i \quad \text{and} \quad \sum_i G_{ij}(P) = P_j \quad . \tag{11}$$

Hence a mating system can also be described as a collection of non-negative functions $G_{ij}(P)$ depending on $P = (P_1,...,P_n)$ and satisfying (11). If a population would follow this mating system, and if at a certain moment the population has a type distribution $P_i = N_i/N[(i = 1,...,n)]$, then the mating system would dictate that the males and females pair up in such a way that there are exactly $NG_{ij}(P)$ pairs where the male is of type i and the female of type j, this for all $i = 1,...,n$ and all $j = 1,...,n$. In other words, a mating system is a recipe for pairing up:

$$G: \quad P = (P_1,...,P_n) \rightarrow (P_{ij}) \quad . \tag{12}$$

For a population following this recipe, we may rewrite (9) as

$$P'_k = \sum_{i,j} G_{ij}(P) f_{ijk} \tag{13a}$$

or, for brevity,

$$P' = TP \quad . \tag{13b}$$

Here T denotes a continuous transformation (provided the functions $G_{ij}(P)$ are continuous in P, as seems reasonable). Starting with an initial type distribution $P^{(0)}$, after ℓ generations we will have the distribution

$$P^{(\ell)} = T^\ell P^{(0)} \tag{14}$$

assuming that all generations would continue following the given mating system. Of special interest would be the distributions P which remain the same from generation to generation, that is, distributions P which satisfy $TP = P$. One can prove that to each (continuous) mating system there corresponds at least one such equilibrium distribution.

Let me now turn to the following question which I found intriguing: what can be said about the collection of all possible equilibrium distributions for all possible mating systems? Let this collection be denoted by Ω. Then a distribution P belongs to Ω precisely when it is an equilibrium distribution for some mating system. If in nature one finds a population having a distribution not belonging to Ω, then one can be sure about the fact that the population is not in equilibrium from generation to generation, even if one knows nothing about the system of mating using by this population.

In view of (9), it is not hard to see that a distribution P (that is, a set of non-negative numbers $P_1,...,P_n$ with $\sum P_i = 1$) belongs to Ω

if and only if one can find non-negative numbers P_{ij} which satisfy not only equation (8), but also

$$\sum_{i,j} P_{ij}f_{ijk} = P_k \qquad\qquad (15a)$$

or, what is the same,

$$\sum_{i,j} N_{ij}f_{ijk} = N_k . \qquad\qquad (15b)$$

It follows that the set Ω is even convex.* Let us see what Ω looks like for our original population with types 1, 2, 3 as the genotypes AA, Aa, aa. The expected proportion $f_{ijk} = f_{jik}$ of children of type k for a couple of type (i,j) as listed in Table 2.

TABLE 2

ij =	11	22	33	23	31	12
f_{ij1}	1	¼	0	0	0	½
f_{ij2}	0	½	0	½	1	½
f_{ij3}	0	¼	1	½	0	0

We claim that in the present case, the set Ω corresponds to the shaded area in Figure 3. Let us first show that at least the shaded region is a subset of Ω. Since Ω is convex, one only needs to consider the five

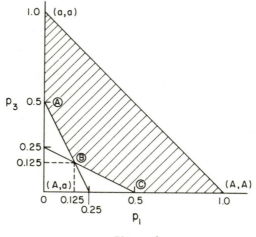

Figure 3

---

*A convex space Ω is a set of points in the Euclidean space of n dimensions such that if x ϵ Ω, y ϵ Ω then also ax + (1 - a) y ϵ Ω, for 0 < a < 1. Geometrically this means that if two points belong to Ω, the line segment joining them also belongs to Ω.

vertices of this shaded region. Thus we must show that each vertex corresponds to an equilibrium distribution for some (appropriately chosen)
system of mating. There is no difficulty for the vertices where
(P_1, P_2, P_3) equals $(1, 0, 0)$ or $(0, 0, 1)$; such homozygous populations
cannot help but copy themselves from generation to generation. It remains
to consider the vertices labeled A, B, and C in Figure 3.

A	P_1	P_2	P_3	B	P_1	P_2	P_3	C	P_1	P_2	P_3
	0	1/2	1/2		1/6	2/3	1/6		1/2	1/2	0
	AA	Aa	aa		AA	Aa	aa		AA	Aa	aa

A mating system which will reproduce a distribution of type A is one
which requires that only Aa and aa can mate; that is, we only allow matings of type (2,3) or (3,2) and forbid matings of type (2,2) or (3,3). A
similar mating system is used for the type C distribution. A mating system which will reproduce a type B distribution from generation to generation is to pass a law that AA mates only with aa and that Aa mates only
with Aa.

It is possible, therefore, to establish the five vertices of Figure
3 as being equilibrium distributions. It is worthy of repetition, however, that the points A, B, and C required the instituting of laws in
order to insure their equilibrium. In society these laws may be of a
legislative form, or they could involve a caste or class structure, or
even the establishing and propagating of prejudice. In nature, however,
these "laws" would necessarily be of a more evolutionary origin and less
from a conscious intent of the members of the population.

Next we must prove that no point outside of the shaded region can
belong to Ω. In other words, we must show that a distribution
(P_1, P_2, P_3) belonging to Ω necessarily satisfies the pair of inequalities

$$4P_1 + 2P_3 \geq 1 \quad ; \quad 2P_1 + 4P_3 \geq 1 \tag{16}$$

(which determine the lower boundary of the shaded region).

It will be convenient to assume that the matrix (P_{ij}) is symmetric,
that is, $P_{ij} = P_{ji}$ for all i and j. This assumption can be made without
loss of generality. Suppose we have a type distribution (P_1, P_2, P_3)
with corresponding possibly non-symmetric matrix (P_{ij}) satisfying (8);
the type distribution of the offspring is then given by equation (9).
Now form the symmetric matrix $P^*_{ij} = (P_{ij} + P_{ji})/2$. One easily verifies
that this symmetric way of pairing up is also admissible for the given
type distribution and, further, that this system of mating results in exactly the same type distribution for the offspring; here, one heavily uses
the fact that $f_{ijk} = f_{jik}$ for all i, j and k.

Consider a type distribution (P_1, P_2, P_3) belonging to Ω and the
associated matrix (P_{ij}) satisfying (8) and (15). In view of the above
remarks, we may assume that $P_{ij} = P_{ji}$. It must be shown that (16) holds.
We write out equation (8) explicitly,

$$P_1 = P_{11} + P_{12} + P_{13} \quad , \tag{8a}$$

$$P_2 = P_{22} + P_{23} + P_{21} \quad , \tag{8b}$$

$$P_3 = P_{31} + P_{32} + P_{33} \quad . \tag{8c}$$

Combining these equalities by multiplying equation (8a) by three, sub-
tracting equation (8b) and adding equation (8c), yields

$$3P_1 - P_2 + P_3 = 3P_{11} + 2P_{12} + 4P_{13} + P_{33} - P_{22} .$$

Hence, using that $P_1 + P_2 + P_3 = 1$, we have

$$4P_1 + 2P_3 - 1 = 3P_{11} + 2P_{12} + 4P_{13} + P_{33} - P_{22} . \tag{A}$$

A similar combination yields,

$$4P_3 + 2P_1 - 1 = P_{11} + 2P_{32} + 3P_{33} + 4P_{31} - P_{22} . \tag{B}$$

We note that in each of these latter two equations, we have the term $4P_{13} - P_{22}$, $(P_{13} = P_{31})$.

Next, we have the equilibrium condition that $P_k = P_k'$ for all k. In particular, using Table 2,

$$0 = P_3' - P_3 = \sum_{i,j} P_{ij}f_{ij3} - \sum_i P_{i3}$$

$$= (\tfrac{1}{4}P_{22} + P_{33} + \tfrac{1}{2}P_{23} + \tfrac{1}{2}P_{32}) - (P_{13} + P_{23} + P_{33})$$

$$= \tfrac{1}{4}P_{22} - P_{13}$$

here we also used that $P_{23} = P_{32}$. It follows from (A) and (B) that, in an equilibrium situation (from generation to generation), we must have

$$4P_1 + 2P_3 - 1 = 3P_{11} + 2P_{12} + P_{33} \geq 0$$

$$4P_3 + 2P_1 - 1 = P_{11} + 2P_{32} + 3P_{33} \geq 0$$

and this implies the inequalities (16).

We have, therefore, shown that the equilibrium distribution for an arbitrary system of mating satisfies (16). Also, that each distribution satisfying (16) can be in equilibrium, that is, belongs to Ω. Consequently, Ω is given by the shaded region in Figure 3.

Let us add another degree of complexity to our problem by lifting the restriction on everyone being mated. That is, we now allow bachelors and spinsters in our model. This will not invalidate our previous results, but does necessitate a generalization of our previous definitions. As before, let N_{ij} denote the number of couples where the male is of the type i and the female of type j, (i = 1,...,n; j = 1,...,n). Further, let

$N_{io} \equiv$ number of unmated males of type i ,

$N_{oj} \equiv$ number of unmated females of type j .

This leads to the following augmented matrix

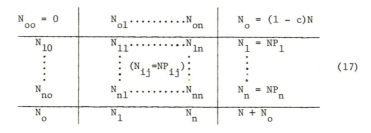

$$\begin{array}{c|cc|c}
N_{oo} = 0 & N_{o1} \cdots\cdots N_{on} & N_o = (1 - c)N \\
\hline
N_{10} & N_{11} \cdots\cdots N_{1n} & N_1 = NP_1 \\
\vdots & \vdots \quad (N_{ij}=NP_{ij}) \quad \vdots & \vdots \\
N_{no} & N_{n1} \cdots\cdots N_{nn} & N_n = NP_n \\
\hline
N_o & N_1 \qquad N_n & N + N_o
\end{array} \tag{17}$$

Here, $N_o = \sum_i N_{io}$ is the total number of unmated males (bachelors); $N_{oo} = 0$ for convenience. The total number of bachelors is equal to the total number $N_o = \sum_j N_{oj}$ of unmated females (spinsters). We put $N_o/N = 1 - c$, so that $(N - N_o)/N = c$ is the proportion of individuals which do mate, (the same for males and females). We shall call c the mating ratio.

Letting $P_{ij} = N_{ij}/N$ (in particular, $N_{io} = NP_{io}$ and $N_{oj} = NP_{oj}$), one sees that the equations (7), (8), (9), and (15) are valid, provided the summations over i or j extend over the set $0,1,\ldots,n$ and provided that we put $f_{ijk} = 0$ when one of i, j, or k equals 0. In (9) and (15), each double sum should be preceded by an additional factor 1/c or instead one may replace (10) by $\sum_k f_{ijk} = 1/3$ (if $1 < i, j < n$). Further, $P_o = N_o/N = 1 - c$.

We shall only be interested in the possibility of unmated males and females when something prohibits the formation of couples of certain specific types (i,j). Implicit in this restriction is the assumption that an individual will mate unless there is a cause for his not doing so. For a human population, such a cause may be custom, prejudice, mores, etc. In lower life forms there would be other causes, for example imprinting which will be discussed below. It is to be recalled that we are discussing large populations and expectations, so that a more correct statement of the above would be that, on the average, an individual will always mate unless there is a cause for his not doing so.

We shall refer to all the causes which prohibit couples of type (i,j) from forming as taboos. In fact, let Γ denote the set of all pairs (i,j) such that

$$(i,j) \ \varepsilon \ \Gamma \Rightarrow N_{ij} = NP_{ij} = 0 \ ,$$

which is a formal statement of the fact that there are no couples ($N_{ij} = 0$) of the type (i,j) if (i,j) is in the taboo set Γ. Two other conditions seem reasonable in the case of such taboos. These are

(Ia) $(i,j) \ \notin \ \Gamma \Rightarrow P_{io}$ or $P_{oj} = 0$,

(Ib) $(i,j) \ \varepsilon \ \Gamma <= P_{io} > 0$ and $P_{oj} > 0$,

and, further,

(IIa) $(i,j) \ \notin \ \Gamma$ and $P_i > 0$ and $P_j > 0 \Rightarrow P_{ij} > 0$,

(IIb) $(i,j) \ \varepsilon \ \Gamma <= P_i > 0$ and $P_j > 0$ and $P_{ij} = 0$.

In words, condition (I) states that when a particular pair (i,j) is not taboo (not in Γ), then this implies that the mating cannot be such that there are both unmated males of type i and unmated females of type j; after all, this would not make sense since there is no taboo against their mating each other. Condition (II) requires that if a particular mating type (i,j) is not taboo and there are males of type i in the population and also females of type j ($P_i > 0$ and $P_j > 0$), then there are at least some matings between these types ($P_{ij} > 0$). Otherwise, (i,j) would amount to a new taboo. Remember that we are thinking of very large populations as noted above.

An example of a type of mating system with taboos was brought to my attention a few years ago by Professor Marvin Seiger, and it was in fact the stimulus for this investigation. He made the observation that certain pigeons appear predominantly in two color tones, light and dark as opposed to a continuum of intermediate shades. He hypothesized that there was

some kind of imprinting mechanism (i.e., a very deep impression made by
the parents on the young during a critical week in their development)
which causes the offspring to select only a mate which has the color of
at least one of its own parents. One taboo in this case would be that
the offspring of two dark parents would never choose a light mate while
the offspring of two light parents would never choose a dark mate. He
further conjectured that under a certain type of random mating (which we
will not specify now) consistent with the above taboos, the population
would become purely homozygous in the long run, (only individuals AA and
aa with parents of the same type). Here it is assumed that the color
characteristic of the pigeon is determined by two genes A and a, where A
is dominant, such that the combinations AA and Aa correspond to a dark
(D) individual and aa to a light (L) individual. His conjecture turned
out to be correct. But let us now ask a different question. Namely,
suppose we allow all possible sorts of mating systems; do there then exist situations where a population can be in equilibrium from generation
to generation without being purely homozygous?

Because the behavior of an individual depends on the complexion of
its parents, we make the following classification for this individual.

1. genotype AA, parents DD
2. genotype Aa, parents DD
3. genotype Aa, parents DL or LD
4. genotype aa, parents DD   TABLE 3
5. genotype aa, parents DL or LD
6. genotype aa, parents LL

The taboos are as follows:

parent	D–D	...	...	L–L
offspring	...	L	D	...
types	[1,2,4]	[4,5,6]	[1,2,3]	[6]
		taboo		taboo

The first type of taboo (against mating a light individual) forbids 17
pairs (i,j), $[17 = 2 \cdot 3 \cdot 3 - 1$, only $(4,4)$ having $i = j]$. The second type of
taboo (against mating a dark individual) forbids 6 pairs among which only
the pairs $(3,6)$ and $(6,3)$ are new. Altogether there are 19 taboo pairs
in the set Γ and only $36 - 19 = 17$ admissible pairs (i,j), which are
listed in Table 4.

TABLE 4

	Total
type 1 can only mate types 1, 2, 3	3
type 2 can only mate types 1, 2, 3	3
type 3 can only mate types 1, 2, 3, 4, 5	5
type 4 can only mate type 3	1
type 5 can only mate types 3, 5, 6	3
type 6 can only mate types 5, 6	2
	17

To clarify how one determines the possible pairs in Table 4, given
the above-mentioned taboo, let us look at the mating possibilities for an
individual in class 4. He is a light individual (genotype aa) with both

parents dark; therefore, he will only want to mate with a dark female.
Thus he will reject the individuals in class 4 or class 5 or class 6 since
they are light and he is interested in mating only with members of the
classes 1, 2 or 3 since they are dark. A reverse phenomenon now occurs
in that he in turn is rejected by those in classes 1 and 2 since he himself is light. His only choice for a mate is thus an individual of
class 3. Note that individuals of class 3 happen to have plenty of other
choices as well (see Table 4).

We wish to find all possible equilibrium solutions (that is, distributions which are members of Ω), given the above-mentioned taboos and
given the above two conditions (I) and (II). Hence, Ω now consists of
the type distributions $P = (P_1, P_2, P_3, P_4, P_5, P_6)$ such that one can find
a (usually non-symmetric) matrix (P_{ij}), i.e., system of mating, with i,
$j = 0,\ldots,6$ $(P_{oo} = 0)$, such that

$$P_{ij} \geq 0 \text{ and } P_{ij} = 0 \qquad \text{if } (i,j) \in \Gamma$$

and, further,

$$P_i = \sum_{j=0}^{6} P_{ij} \quad ; \quad P_j = \sum_{i=0}^{6} P_{ij} \quad ; \tag{18a}$$

$$P_k = 1/c \sum_{i=1}^{6} \sum_{j=1}^{6} P_{ij}f_{ijk} , \tag{18b}$$

where

$$c = \sum_{i,j=1}^{6} P_{ij} . \tag{19a}$$

Thus,

$$1 - c = \sum_{i=1}^{6} P_{io} = \sum_{j=1}^{6} P_{oj} . \tag{19b}$$

Finally, we insist on the conditions (I) and (II).

Since it is a fraction, c of the population that has any offspring
[equation (19)], we can stabilize the population size by letting each
couple on the average have $1/c$ male and $1/c$ female offspring, see (18b).
The offspring types are determined by the frequency coefficients f_{ijk}
which are easily computed.

Because a mating of a type 6 male with a type 6 female is not taboo
$[(6,6) \notin \Gamma]$, it follows from rule (Ia) that either there are no bachelors
of type 6 $(P_{60} = 0)$ or there are no spinsters of type 6 $(P_{06} = 0)$. For
concreteness, let us assume that $P_{60} = 0$ so that all the males of type 6
get mated; (the argument can be done with equal validity assuming that all
the type 6 females are mated). We see, by inspection of the genotypes in
Table 3 that all the offspring of couples of types (5,5), (5,6), (6,5),
and (6,6) will be of type 6. In particular, the NP_6 males of type 6 be-
get $NP_6(1/c)$ male offspring all of type 6. In order that in the filial
population type 6 individuals are no more numerous than in the parent
population, we must have either $P_6 = 0$ or $c = 1$; in either case there
would be no unmated males or females of type 6, $(P_{60} = P_{06} = 0)$. Further,
$(P_{55} + P_{56} + P_{65} + P_{66})(1/c) = P_6 = P_{65} + P_{66} = P_{56} + P_{66}$, and this immediately yields that

$$P_{55} = P_{56} = P_{65} = 0.$$

Since the mating of type 5 males to type 5 females is not taboo (see
Table 4), it would follow from $P_{55} = 0$ and rule (II) that $P_5 = 0$, i.e.,
there are no individuals of type 5. Now type 3 and 5 individuals are only
produced by couples of types (3,4) or (3,5) and in equal numbers so that,
in any equilibrium, $P_3 = P_5$. This assertion may be validated by looking
at Table 3. Here we see that a (DL type) couple (3,4) is an Aa-type mating with an aa-type, so that one has both Aa (type 3) and aa (type 5)
offspring with equal frequency. Similarly, for a (3,5) couple. Since
$P_3 = P_5$, and $P_5 = 0$, we have $P_3 = 0$.

But now, since there are no individuals of type 3 and type 4 individuals can mate only with type 3, the males and females of type 4 (which
are equal in number) have no choice but to remain unmated. They are all
of genotype aa and thus one would have a drain of the type a genes, at
least when $P_4 > 0$. This would have to be compensated by a corresponding
draining away of A genes caused by unmated males and females of one of
the types 1 or 2. Since the total number of unmated males must equal the
total number of unmated females, there would be an equal number of unmated
males of types 1 or 2 as unmated females of types 1 or 2. But this would
violate rule (I) since males and females of the types 1 or 2 can mate each
other [that is, there are no taboos (1,1), (1,2), (2,1) or (2,2)].

We conclude that necessarily $P_4 = 0$. Now couples of type (2,2)
would have some offspring of type 4; therefore $P_{22} = 0$. But (2,2) is not
taboo; therefore $P_2 = 0$.

Summarizing, we have shown that there can be only the trivial equilibrium situation where $c = 1$ and $P_2 = P_3 = P_4 = P_5 = 0$; thus, $P_1 + P_6 = 1$.

Also, in any dynamic situation approaching equilibrium we must have
a convergence to this purely homozygous population; thus, all hetero-
zygotes must die out.

This is not always what one finds in nature, so the above result
suggests that our model may have to be modified. A particularly realistic
situation would be the one in which only the males imprint while the females are free from taboos. I have worked out this case but the results
are too lengthy to present here. I will merely say that there exist non-
trivial equilibrium situations for this case [which do satisfy both conditions (I) and (II)] as opposed to the trivial solution $P_1 + P_6 = 1$
found above. Unless $P_1 + P_6 = 1$, the mating ratio is always less than
unity; more precisely,

$$\frac{1}{2} + \frac{1}{4} \sqrt{2} \le c < 1 \, ,$$

and these bounds cannot be improved. Thus, at most 15% of the population
remains unmated in any equilibrium situation.

As a final example, which takes us from genetics to ecology, let us
consider a possible interaction between two types of plants and three
species of animals.

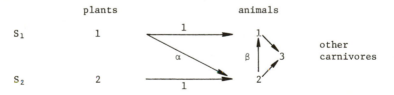

The arrows indicate the direction of energy flow. The numbers 1, 1, α
and β at the arrows give the food value of one unit to the animal eating
it; these food values are taken relative to the minimal daily requirements.

Thus one animal of type 1 could live on a diet of 1 unit of plant 1 per day and also on a diet of $1/\beta$ animals of type 2 per day, or on some convex mixture of these two diets. Further, let the ith plant grow by a given amount S_i per day. Finally, let the reproduction rate of the jth animal be equal to the given number ρ_j.

Consider an equilibrium situation where there are N_i animals of type i.

Suppose the animals of type j eat, on the average, A_{ij} animals of type i per day, and an amount B_{ij} of plant i per day. This leads to the necessary relations:

$$\rho_1 N_1 = A_{13} \qquad\qquad\qquad S_1 \geq B_{11} + B_{12}$$

$$\rho_2 N_2 = A_{21} + A_{23} \geq A_{21} \qquad S_2 \geq B_{22} \qquad\qquad (20)$$

$$N_1 \leq B_{11} + \beta A_{21} \quad ; \quad N_2 \leq \alpha B_{12} + B_{22} \ .$$

Thus a pair (N_1, N_2) of non-negative numbers represents possible population sizes for the animals of types 1 and 2 if and only if one can find non-negative numbers A_{ij} and B_{ij} in such a way that all the above inequalities hold. We shall be interested in the set of possible pairs (N_1, N_2).

The system (20) can be simplified considerably. The relation $\rho_1 N_1 = A_{13}$ may as well be omitted since it merely serves as a definition of A_{13}. The second relation serves as a definition of A_{23} provided we retain the inequality $\rho_2 N_2 \geq A_{21}$, and afterward this inequality may also be omitted after writing the third relation (20) as $N_1 \leq B_{11} + \beta \rho_2 N_2$. Again $N_2 \leq \alpha B_{12} + B_{22}$ together with $B_{22} \leq S_2$ may be replaced by $N_2 \leq \alpha B_{12} + S_2$. This leads to the three inequalities

$$N_1 \leq B_{11} + \beta \rho_2 N_2 \qquad\qquad\qquad\qquad\qquad (21)$$

$$N_2 \leq \alpha B_{12} + S_2 \qquad\qquad\qquad\qquad\qquad\qquad (22)$$

$$B_{11} + B_{12} \leq S_1 \ . \qquad\qquad\qquad\qquad\qquad\qquad (23)$$

In particular, $B_{11} = 0$ is only possible when $N_1 \leq \beta \rho_2 N_2$ while $B_{12} = 0$ is only possible when $N_2 \leq S_2$. Eliminating the non-negative unknowns B_{11}, B_{12}, we finally obtain the following necessary and sufficient conditions for (N_1, N_2) to belong to equilibrium population sizes:

$$\max(0, \ N_1 - \beta \rho_2 N_2) + \max[0, (N_2 - S_2)/\alpha] \leq S_1 \ .$$

This is equivalent to the three inequalities

$$N_2 \leq \alpha S_1 + S_2$$

$$N_1 \leq S_1 + \beta \rho_2 N_2$$

$$N_1 + \frac{1}{\alpha} N_2 \leq S_1 + \frac{1}{\alpha} S_2 + \beta \rho_2 N_2 \ .$$

We may now obtain a graphical solution to the set of equations (21) - (23) as shown in Figure 4 by determining the extreme points of the system.

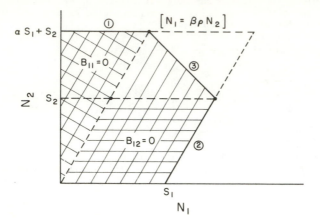

Figure 4

The border labeled (1) in Figure 4 is obtained from inequality (21). The cross-hatched region below corresponds to the choice $B_{11} = 0$. Similarly, the border labeled (2) is obtained from equation (22). The choice $B_{12} = 0$ leads to the subset $N_2 \leq S_2$. The third border (3) is obtained from equation (23). Its slope is positive, precisely when $\beta \rho_2 \alpha > 1$. In that case, it is more efficient for the animals of type 1 to allow the animals of type 2 to eat the plants of type 1, since the resulting food value (by eating the animals of type 2) exceeds the food value derived by eating the plants of type 1 directly. An investigation is presently under way for the more general case of animals $1, 2, \ldots, m$ and plants $1, 2, \ldots, n$. The system inequalities is here $A_{ij} \geq 0$, $B_{ij} \geq 0$ and, further,

$$N_j \leq \sum_i A_{ij} a_{ij} + \sum_i B_{ij} b_{ij} \ ,$$

$$S_i \geq \sum_j B_{ij} \ ,$$

$$\rho_i N_i = \sum_j A_{ij} \ .$$

It is also possible to incorporate the previously established notion of a taboo by requiring $a_{ij} = 0$ or $b_{ij} = 0$ for certain pairs (i,j). This should prove an interesting course for future research.

REFERENCES

1. J. H. B. Kemperman, On Systems of Mating, *Indagationes Mathematicae* 29 (1967), I(245-261); II(262-276); III(277-290); IV(291-304).
2. M. B. Seiger, A Computer Simulation of the Influence of Imprinting on Population Structure, *The American Naturalist*.

A MATHEMATICAL THEORY OF POLITICAL COALITIONS

William Riker

University of Rochester

Model building may proceed by a number of different paths, none of which is intrinsically superior. Consider the discussion by Professor Kemperman. He reviewed a relatively simple natural system which could be easily described by a simple mathematical model. A more complex system was then considered with a correspondingly more complex model. The explicit assumptions made in this latter model were then relaxed in a series of steps and at each step the solution to the model was compared to a different natural system. The important characteristic about such model formulation is that one is directly describing a natural system in mathematical terms.

A second type of model construction, and the one which we will be concerned with here, is the axiomatic method. This method, as implied by its name, involves selecting a number of axioms or given assumptions and from these constructing a mathematical theory. The mathematical theory is then compared with the natural world. If the correspondence is close, one has confidence that the assumptions are in fact appropriate and the description, thus indirectly arrived at, is approximately correct. If the correspondence is not close, however, reconsideration is necessary.

The former method produces simply a description but, if one has confidence in the axioms, the latter also provides an explanation, a necessary and sufficient condition. One can say that, given the axioms, the behavior described is necessary because it is what must logically occur in a system with the given characteristics. Furthermore, since the axioms are presumably descriptive, they are sufficient logically to bring about the observed behavior in the natural system.

It is often the case that a given set of problems and appropriate axioms requires a specific kind of mathematics for its solution. For example, it was a necessary condition for the development of classical mechanics in physics that Newton "invent" the differential calculus. The discipline of "game theory" was developed for just such a purpose. We quote from *Theory of Games and Economic Behavior* by Von Neumann and Morgenstern:

> The importance of the social phenomena, the wealth and multiplicity of their manifestation, and the complexity of their structure, are at least equal to those in physics. It is, therefore, to be expected--or feared--that mathematical discoveries of a stature comparable to that of calculus will be needed in order to produce decisive successes in this field.

The application of game theory by the above authors was in the field of economics but, because of the generality of the mathematical structure,

application in other fields in the social sciences is also possible. The
following discussion concerns one such application in the study of politics.
     One might think of a game as a tree or a branching network, which
describes all possible ways that a game might be played and might end:

Move Number	Player Number
.	.
.	.
.	.
4	2
3	1
2	2
1	1

At each node a player (of whom there may be n, though only 2 in the example) has an opportunity for choice among alternative actions which are
represented as branches from the node (e.g., a and b at move 1). These
alternatives are a set of all actions relative to the game as described
in the rules and not prohibited by the rules (e.g., in chess, at the first
move, the sixteen possible plays of pawns and the four possible plays of
knights). The choice by a player at move k-1 determines the node for the
next player at move k. Thus, in the diagram, whether player 2 is at node
A or node B at move 2 depends on whether player 1 chose alternative a or
alternative b on the previous move. The history of a particular match
under the rules is a path up the tree to a particular outcome $(\alpha_1, \alpha_2, \ldots, \alpha_s)$ from move r. Thus all possibilities of action in a game are strictly
derived from and implicit in the rules. In that sense a game and its
rules are equivalent.
     The game theoretic model allows one thus to study the rules (i.e.,
game) in the abstract apart from the historical context and the psyches
of individual participants. We can postulate that the participants (i.e.,
players) are rational in the sense of desiring to maximize expected utility from the play. Then we can construct a normative theory about the
best way to play. Insofar as the postulate of rationality is descriptive
of actual human behavior, the theory is then also just as descriptive as
it is normative.
     Because the interest is in the interpretation of the model in the
context of political science and not in the mathematics, *per se*, we will
indicate how rather straightforward observations in the theory of games
can be used to draw inferences in the social sciences. The present state
of the art allows one to make general relational statements, as opposed
to detailed particular statements. This may or may not be considered a
weakness of the model, dependent upon one's mathematical predilections.
The mathematics will be interlaced with interpretation and the major in-
ference, i.e., the size principle, will be extensively supported by historical examples. In essence, the size principle is a statement of how
large a political coalition will become under certain specified conditions
within the model. A more detailed description of this principle will be
deferred until the basic game theoretic notions are developed.
     To introduce the basic concepts of game theory as we will use them,
let us consider a group of n individuals. Assume that this group as a
whole is faced with a problem which requires from them a yes or no decision. The group iteslf has an intrinsic value in that it has the power
of bringing about a decision and of implementing it. Each member of the
group also has an individual value, i.e., his particular vote, his ability
to influence or coerce others, etc. The rules of decision making, i.e.,
the game, determine the ways in which the group as a whole can reach a
decision on the problem. Assuming that the game is to decide by voting
the individuals can, at most, form three subgroups within the group:

those that vote yes, those that vote no, and those that abstain from voting.

These subgroups also have a value, that is, the ability to determine the decision of the group. Further, each member of the subgroup may be allocated a certain portion of the total value of the coalition. This portion may take the form of a promise for future action (on policy, on votes, etc.) or it may take the form of money, etc.

If members of the group have equally weighted votes and if the final decision is based on simple majority voting, then any subgroup of size $[(\frac{n}{2} + 1; n \text{ even})]$ controls the decision. If such a subgroup can be formed privately before the vote is registered, and if the members of this subgroup may keep others from joining their ranks, then the members of this subgroup may control the whole value in the game. If, however, the votes are taken in secret, then those in a previously formed winning subgroup will not be able to keep others from joining it. This is essentially the "bandwagon" effect, where everyone wants to be on the winning side.

The restricted example above serves to illustrate a few of the fundamental concepts with which we will be dealing. Initially we have a group of individuals that plays a game (decision making process), the group as a whole having a certain value. The members of the group, each of which has a value, form and join coalitions (subgroups) so as to change imputations (i.e., lists of pay-offs to players) to their advantage. Individual players may change from one coalition to another so as to receive the largest possible pay-off in an imputation. Finally, there exists some minimum size for a coalition which is necessary to win.

Recalling the above example and concepts, let us construct our mathematical model. We restrict our analysis to those games in which coalitions have control over who becomes a member. This restriction occurs often enough in nature so as to make its analysis worthwhile. Even a legislature (where members are free to vote as they choose) may have this feature if the original members on the winning side write a bill that is unpalatable to everyone not in their coalition.

The type of theory we will be discussing is zero-sum n-person game theory, which was developed in the works of Von Neumann and Morgenstern. We want to determine how, through the formation of coalitions, an individual may maximize the value of the coalition in which he is a member, as well as his imputation from that coalition. The quantifier "zero-sum" means that the total value of the game is derived, without remainder, from those playing the game. For example, if we have two persons playing the game, then what is won by one person is exactly what is lost by the other person.

Consider a coalition, which we denote by the symbol S, the value of which may be described through a function v(S). The function v(S) describes the rules of the game, since it takes on a value for each of the possible coalitions formed. The importance of v(S) is self-evident since it indicates which of a number of possible coalitions is preferable. Consider the coalitions $S_1, S_2, \ldots, S_n$ the characteristic function of which is $v(S_1), v(S_2), \ldots, v(S_n)$ and these values are ordered such that $v(S_1) \geq v(S_2) \geq \ldots \geq v(S_n)$. In this case the value of the coalition S_1 is greater than or equal to that of S_2, which is greater than or equal to that of S_3, etc. Neglecting other considerations for the moment, a reasonable person orders his preferences over the coalitions in the same manner as the value of the coalitions are ordered. He prefers to be a member of S_1 to being a member of S_2, to S_2 rather than S_3, etc.

The function v(S) is, therefore, the quantity with which we are particularly interested. For large and complicated games, the detailed form of this function is prohibitively difficult to obtain, but we may determine its possible range of values by examining its values in certain extreme cases. Consider the coalition which has no members; we denote the set of such no-member coalition by $\{\phi\}$. The value of this coaltion S = $\{\phi\}$ is,

$$v(\{\phi\}) = 0 \quad , \tag{1}$$

that is, the value of the coalition which has no members is zero.

Another property we desire in the characteristic function is that the value of two coalitions increases when they combine. We, therefore, have the superadditive property that the union of two coalitions S and T has a value greater than or equal to the sum of the values of the two coalitions separately,

$$v(SUT) \geq v(S) + v(T) \quad , \tag{2}$$

where SUT denotes the coalition (union) of the two coalitions S and T. If this condition did not hold, there would be no motivation for small groups (or individuals) to join together into larger groups. Hence, nothing would happen in this game.

Consider the case where we have n-individuals who have formed a large coalition S and a number of smaller coaltions P, Q, R ... . Because of the zero-sum nature of the game, there is a predetermined value that the coalition S may have,

$$v(S) = - v(-S) \quad . \tag{3}$$

In words this means that if one has formed a given coalition S of m-members with a value $v(S)$, then the value of a complementary coalition $-S$ of (n-m)-members, whether formed or not, is equal to the negative of the value of the given coalition. Further, the *best* the smaller coalitions can do, to either minimize their losses or maximize their gains, is to form this complementary coalition, i.e., $-S = P \cup Q \cup R \cup ...$. This is due to the fact that the sum of everything that is gained (lost) by one coalition S is equal to the sum of everything that is lost (gained) by its complementary coalition $-S$.

A consequence of conditions (1) and (3) is that the coalition of which everyone is a member ($S = I_n$) has a zero value,

$$v(\{I_n\}) = 0 \tag{4}$$

The implication is that since there is no losing coalition, then the zero-sum feature of the game states that there is nothing to be won, that is,

$$\overset{(3)}{v(\{I_n\})} = -v(-\{I_n\}) \overset{(1)}{=} -v(\{\phi\}) = 0$$

since $\{\phi\}$ is the complementary coalition of $\{I_n\}$.

The final restriction on $v(S)$ concerns a player who is left alone in the game, that is, who is not a member of a larger coalition. The value of this coalition, of the ith individual alone, is

$$v(\{i\}) = - \gamma \tag{5}$$

where γ is an arbitrary positive constant.

Such a function [v(S)] as defined by conditions (1) through (5) purports to give a complete description of the rules of the game. Once the
above five characteristics of v(S) have been established, one has ex-
hausted his knowledge without making additional assumptions. We may use
these five conditions and construct a graph indicating the range of possible value of v(S), (see Von Neumann and Morgenstern). Figure 1 is such

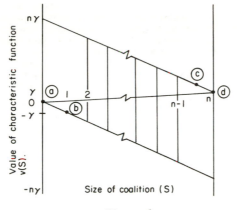

Figure 1

a graph and plots the value of a coalition [v(S)] vs. the size of the coalition. One uses the conditions that the value of the coalition of all
players and no players [v({I_n}) and v({φ})] both have value zero to determine the points (a) and (d) in Figure 1. The point (b) expresses condition (5), i.e., the value of a coalition of a single individual is -γ.
The final point drawn (c), may be obtained by using conditions (3) and
(5), that is, the coalition of (n-1) individuals has a value,

$$v(\{n-1\}) = -v(-\{n-1\})$$

but -{n-1} = {1}, so that v({n-1}) has the value +γ.
The straight line connecting the points (a), (b), and (-n, n) and
the straight line connecting the points (n, 0), (c) and (d) determine the
extreme values of the characteristic function v(S). Therefore, the value
[v(S)] of a given coalition of m-persons, out of a possible n-persons,
may have the values $-m\gamma \leq v(S) \leq (n-m)\gamma$. This implies that a coalition
of m-persons (S) can at most win what is lost by the coalition of the remaining (n-m)-persons [(n-m)γ], or at worst lose the complete value of
the coalition, [-mγ].
The graph of the characteristic function in Figure 1 does not completely describe the n-person game, however. What is of importance to
the individual player, in addition to the value of the coalition, is the
portion of that value which he personally receives. It is quite possible
that an individual j, whose receipts we denote by x_j, may prefer a coalition S_2 to S_1 even though $v(S_1) > v(S_2)$. If the payment to j is greater
when he is a member of S_2 than when he is a member of S_1, he will prefer
S_2 to S_1 even though the value of S_1 is greater than that of S_2. One
must, therefore, not only describe the pay-off to the coalition, but also
the individual imputations. The two following conditions describe the
characteristics of the imputations:

$$x_j \geq v(\{j\}) \tag{6}$$

and

$$\sum_{j=1}^{m} x_j = 0 \tag{7}$$

Condition (6) asserts that no individual will accept in any coalition an imputation less than the value he had in a coalition of himself, alone. Condition (7), in addition to giving the zero-sum feature of the game, also asserts that rational players, whatever their coalition structure, will obtain the full value of the game.

The range of values of v(S) in its present form (Figure 1) is not of much use to us. So, to restrict this range further, we introduce the notion of a majority. Let m be the minimal number of individuals in a majority so that, out of a group of n individuals,

$$[n] \leq m \leq n$$

where

$$[n] = \begin{cases} \dfrac{n+1}{2} & \text{for n odd,} \\[2ex] \dfrac{n}{2} + 1 & \text{for n even.} \end{cases}$$

It is now useful to introduce an assumption from the social sciences to restrict the range of v(S). Let us assume that coalitions below a certain size are losers, and that those at and beyond a certain size are winners. This assumption implies that there exist certain minimal winning coalitions in social situations. We can, therefore, use this notion of a majority to separate the possible coalitions into three sets: W ≡ the set of winning coalitions, B ≡ the set of blocking coalitions, and L ≡ the set of losing coalitions.

We can define the three sets W, B, and L, as follows: If a coalition $[S_p]$ has p members and p is greater than the number of individuals not in S_p, and p is also greater than or equal to that required for a majority (m), then S_p is a member of the set of winning coalitions $[S_p \in W]$. Further, if S_p is a member of the set of winning coalitions but is not a grand coalition, i.e., p ≠ n, then the value of the coalition S_p is positive $[v(S_p) > 0]$. We also note that if p is exactly equal to the majority size (m), then $S_p \in W^m$, where W^m is the set of minimal winning coalitions; that is, if $S_p \in W^m$, then $(S_p-1) \notin W$. The set W^m contains all coalitions that have the minimum number of members necessary to win; i.e., if S_p is reduced by one member, it is no longer a member of W^m.

If a coalition has q members, where q is less than that necessary for a majority but is greater than or equal to the number of remaining individuals in the game, then S_q is a member of the set of blocking coalition $[S_q \in B]$. This set contains all coalitions that are of such a size as to prevent any coalition from winning. The value of such a coalition is zero $[v(S_q) = 0]$ since, by the zero-sum condition, if no one wins, no one loses.

If a coalition is not in the set of winners $[S \notin W]$ and it is not in the set of blocking coalitions $[S \notin B]$, then we say it is a member of the set of losing conditions $[S \in L]$. If the number of people (p) in this coalition is greater than zero (p > 0), and given that the value of a winning coalition is positive, then we have from the zero-sum condition that the value of a losing coalition is negative, i.e., if $S \epsilon L$ then $v(S) < 0$.

The above three conditions may be summarized in the following three inequalities,

$$S_p \epsilon W \Rightarrow 0 < v(S) < (n - p)\gamma \qquad (8)$$

$$S_p \epsilon B \Rightarrow v(S) = 0 \qquad (9)$$

$$S_p \epsilon L \Rightarrow -p\gamma < v(S) < 0 \qquad (10)$$

where the coalition S_p has p members out of a total of n-persons in the game. These inequalities further restrict the range of $v(S)$ and are depicted in Figure 2 where we have ignored the blocking coalitions. We have thus narrowed the range of the characteristic function, but we still know relatively little about what coalitions might occur and about what imputations might be associated with them.

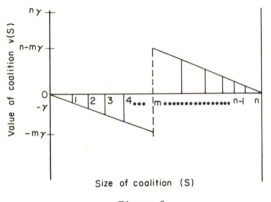

Figure 2

We will now state the result, [for a proof, see reference (2)]: if there exists some S_q, $S_q \epsilon W$, and some imputation $(x_1,...,x_m)$ associated with S_q such that S_q can guarantee its members more than they might receive in a larger one, then they would prefer coalitions of size q to all others. Such coalitions S_q are realizable, while all others S_p, $p \neq q$, are unrealizable. Presumably, once a coalition reaches a realizable size it is relatively stable except, of course, for internal squabbles over the division of $v(S)$ into x_j, $j \epsilon S$.

The notion of realizable coalitions is that, in the set W, there is a subset of realizable coalitions W^q such that any coalition in W^q is preferred to any coalition not in it because S, $S \epsilon W^q$, the amounts that S can unilaterally (that is, without the cooperation of -S) guarantee its members individually are at a maximum and, for that maximum, the costs of organization are minimal. Thus only minimal willing coalitions are realizable. That is, $W^q = W^m$.

Given the range of coalitions (S) indicated in Figure 1 and the restricted range of the characteristic function [v(S)] curve for winning and losing coalitions indicated in Figure 2, one wishes to know which class of curves rational players will accept. We therefore look at the three curves of Figure 3 as possible forms for $v(S)$, for the winning coalitions.

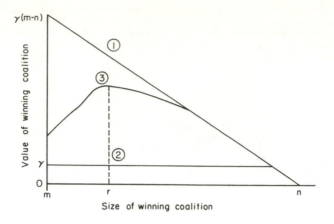

Figure 3

(1) Beginning with the coalition of the whole, which has a value zero,
we see that by reducing the size of the coalition one increases the value
of that coalition. Another effect is that, as members are expelled from
the coalition, their share of the value is distributed among the remaining members. So that there are two effects which provide an incentive to
reduce the size of the coalition.

(2) On this curve there is a range of zero slope which implies that one
can either increase or decrease the size of the coalition without significantly changing the value of the coalition. This implies that there is
less to be divided among the members of the larger coalition, so that the
members of S would expel members until they have the minimal size at the
maximum value.

(3) While this curve is well-defined mathematically and seems to represent a situation in which players have no incentive to reduce the size of
the coalition below the maximum at r, this is not a reasonable social
situation. It is a situation in which, even though a coalition's members
know it is winning, they continue to add to it beyond the point of winning.
It seems unlikely that any situations in the natural world have this feature, especially if it is possible for members of a coalition to control
who enters it. One possibility is to ignore this situation then and to
concern ourselves only with curves (1) and (2). Another possibility is
to examine payments to individuals in the range where the slope of curve
(3) is positive to show that in that range players would prefer a coalition of minimal size. Comparing the gain i.e., to a coalition from adding a member, i, where

$$g_i = v(S + i) - v(S), \quad g_i \text{ positive}$$

with the payment, x_i, to the member added, we can see three possibilities:

   3.1 $x_i > g_i$. Here players already in a coalition would not seek to
expand beyond minimal winning size because it would cost them to do so.
That is, they would have to make up out of their own pockets the amount
of x_i over and above the value of g_i.

   3.2 $x_i = g_i$. There is no motive to expand for the added player ab-
sorbs all the value he brings.

   3.3 $x_i < g_i$. Here there is a motive for old members to bring in a
new member because the new member increases the value of the coalition

more than he costs it. As it turns out, however, it is always possible
for the prospective losers (that is, {S+i} or the complement of {S+i}) to
pay the prospective new member of the winning coalition more to stay with
them (the losers) than to go with the winners. That is, since the game
is zero-sum, the gain from the addition is an increased loss to the remaining losers. If the winners keep some of the additional value, g_i,
generated by the new members for themselves, then it is possible for the
losers to pay that marginal member up to the entire amount, g_i, and the
winners then have no motive to expand. Hence even in this case the expectation is that minimal winning coalitions occur provided, of course,
that players have a large amount of time for negotiation.

The social statement equivalent to the above formal argument is:
"in social situations similar to n-person, zero-sum games with side-
payments, participants create coalitions just as large as they believe
will ensure winning and no larger."

For the first type of example, consider a situation of total war
where there are two coalitions, each with the professed purpose of de-
stroying the government of the opposition. One may consider this contin-
uing process as the game: e.g., the Napoleonic Wars, WWI and WWII were
all a part of an ongoing process; or one may consider the start of a game
as the beginning of a war situation and the end of the game being the
termination of the War.

Now, based on the model, at the end of a war one has a single coalition. That is, the vanquished are now under the control of the winners. This corresponds to the extreme right point (n) in Figure 3 which
has a zero value, i.e., $v(\{I_n\}) = 0$. One would therefore expect that the
wartime coalition would have to dissolve so that the value of the coalition would be enhanced. This model then would predict the failure of
peace markers to maintain the wartime world government, e.g., the Concert
of Europe after the Napoleonic Wars, the League of Nations after WWI and
the United Nations since WWII. All three of these agencies were supposed
to stabilize world politics by perpetuating the coalition of the victors.
All three failed in that respect. The point is that the wartime coalitions had no value after the war had ended.

The second type of example can be taken from American politics
where, on three occasions after a national election, there has remained
only a single party in existence. Consider the election of 1816 after
which the Federalist party almost completely disappeared, leaving the
Republican party as the grand coalition in the "era of good feeling."
This grand coalition was promptly broken up internally before the election of 1824, and even more by the policies of Jackson after 1828.

The interpretation of $v(\{I_m\}) = 0$ in this case reflects the condition
that, in trying to please all the members of a grand coalition, the government cannot pass any controversial legislation. Also, there is no
real control in such a group because the exercise of control always leads
to some members divorcing themselves from the coalition. Therefore, when
Jackson was elected in 1828, he began to reduce the party to a minimal
winning coalition. This he did by expelling lukewarm colleagues and
forming a loyal kernel group.

The Jacksonian coalition, intentionally of a minimum size necessary
to win, eventually destroyed all opposition. This destruction of the opposition led to another grand coalition in 1853-1855, this time of the
Democratic party and, just as in the case of the old Republicans in the
1820's, they were impotent. Unlike the 1820's, however, no national
leader appeared to reduce the coalition; rather, it was reduced by internal splitting. Two centers in the party formed; one around Douglas who
oriented the party around the mid-West (frontier) and the other centered
around Buchanan, who was oriented around the South. The reduction was,

however, too great and the Northern Democrats who flocked to the new
Republican party transformed that losing coalition into a winning
coalition.

A third type of example is a colonial country where the political
structure has essentially two members; those loyal to the colonial government, and the anti-colonials. When this country gains independence,
in general the anti-colonials come into control and form a grand coalition, since the loyalists have been discredited. Almost immediately the
grand coalition, so strongly unified while out of power, begins to reduce
itself in accordance with our model. An immediate example of this would
be India which, after obtaining its independence in 1948, had only the
Congress party. By 1951, however, the candidates for the Congress party
had only 45 percent of the popular vote, but had won 75 percent of the
seats in Parliament. Clearly a minimal winning coalition.

To summarize: Society in general corresponds to a non-zero sum
game; that is, one does not necessarily find that winners win exactly
what losers lose, but there are some clear-cut examples of nearly zero-
sum games, e.g., elections, total wars, etc. Given that zero-sum situations can exist, one may then establish that rational men attempt to form
coalitions that are no larger than are required to win. Also that, in
finding themselves in a coalition not of this size, they will attempt to
reduce or increase their number so as to obtain this optimum size.

REFERENCES

1. J. Von Neumann and O. Morgenstern, *Theory of Games and Economic
    Behavior*, 2nd Edition (Princeton, 1947).

THE PSYCHOLOGY OF SPECULATION: A SIMPLE MODEL*

Bruce J. West

La Jolla Institute
La Jolla, California

INTRODUCTION

The problem we wish to explore in this seminar is the effect of speculation on the stability of the stock market. The present approach differs from that of previous investigators in that we construct a normative model of the stock market which describes the behavior of speculators necessary for both stable and unstable market situations. The model is dynamic and assumes there to be a well-defined point of market equilibrium. A stable market approaches this point as time increases; an unstable market does not. Our model differs from traditional economic models in that the motivation for its general structure is drawn from psychology rather than from the academic notion of the "rational" economic man.

To construct any model one must first determine the salient variables for the process under investigation. To determine the variables of the stock market with which to construct our model, we will give a brief description of how the stock market operates. The operation of the market depends on there being a buyer for a particular stock whenever there is a seller. The buyers and sellers of stock rarely meet since their business is conducted on the floor of the market through the hands of a floor "specialist." There are 360 such specialists on the floor of the New York Stock Exchange (NYSE), each of which handles the buying and selling of a few of the more than 1,000 stocks in the Exchange. It is the function of these experts to determine the present price levels of their individual stock by matching the various buy and sell orders. The matching is generally done by grouping the orders into "lots" of 100 shares each and then matching lots. In doing this matching the floor specialist is purported to stabilize the market; that is, to keep large fluctuations in the price of the stock from occurring. This description regards the function of the specialist to be passive; that is, he does not himself initiate activity, but is merely the steward to the economic forces in the Exchange.[1]

The number of buy (M_B) and sell (M_S) orders (lots) present in the market are generally a function of the present or immediate past price level of the stock. The specialist must determine a price P* such that the difference

*Supported in part by National Science Foundation Grant #5-28501.

$$\Delta M = \left| M_B(P^*) - M_S(P^*) \right| \tag{1.1}$$

is always a minimum. In matching these orders he may activate limit orders on his book, or draw from his own inventory of stock, whichever moves the price more smoothly. Of course, M_B and M_S are also functions of time so that P^* will trace a curve in time for which ΔM is always a minimum (in theory).

The number of orders for a particular stock, as well as the price of that stock, define a set of variables with which a model of the stock market may be constructed. These variables are not unique in that respect, however. Two other variables which are at least as important as the number of buy and sell orders in the market are the number of buyers (N_B) and sellers (N_S) in the market. The introduction of these variables is necessary because the functional dependence of orders and people is not the same. The orders themselves, in the hands of the specialist, are functions of the price. The placement of these orders, however, has more to do with the profit foreseen by the buyers and sellers than with the immediate price level of the stock.

To be able to use the number of buyers and sellers as variables in our model, it is necessary to define a variable of the market to which investors are sensitive. To this end, Osborne[2] analyzed data and determined that the change in the price of stock was not what stimulated investors, but, rather, the percentage change in the price of stock. The motivation for his argument was the Weber-Fechner law in psychology which states that equal ratios of physical stimulus correspond to equal intervals of *subjective* sensation, e.g., equal ratios of light intensity in watts per unit area correspond to equal intervals of brightness. The market variable to be defined is the profit accrued by a particular stock. If $P_j(t)$ is the price of the jth stock at time t and $P_j(t+\tau)$ is the price at time $t+\tau$, then the profit in the interval of time τ is defined as

$$y_j(\tau) = \log_e \frac{P_j(t+\tau)}{P_j(t)} \tag{1.2}$$

Because the quantity $y_j(\tau)$ is sensitive to only percentage changes in the price and not the absolute value of the price, it seems a more reasonable measure of profit than the difference $[P_j(t+\tau) - P_j(t)]$.

To make the choice of variable as given by equation (1.2) more transparent, let us consider the following example. Suppose there are two men, each of whom has a quantity of money to invest. Let us say the first man has \$1,000 with which he buys ten shares of stock A. The second man, who has \$100, buys ten shares of stock B. Let us further assume that both stocks increase in value in such a way that both men realize \$100 return of their respective investments. The first man, therefore, realizes a 10% return on his investment while the second man realized a 100% return. The profit as defined by equation (1.2) would yield

$$y_1 = \log\left(\frac{1100}{1000}\right) = \log(1.1)$$

$$y_2 = \log\left(\frac{200}{100}\right) = \log(2)$$

$$y_2 > y_1$$

so that, although both men made the same number of dollars, the second man made the larger profit.

We have argued above that the profit as defined by equation (1.2) is what stimulates spectators to buy and to sell and, therefore, determined

the number of buyers and sellers within our market. The classical notion
of supply and demand, however, prompts us to look at the relative number
of buyers and sellers instead of N_B and N_S separately. Further, the total
number of speculators in the market ($N = N_B + N_S$) is continually changing,
but this does not necessarily change the pressure to raise or to lower
the price of the stock. The quantity of interest for our model is the
difference between the number of buyers and sellers ($x = N_B - N_S$), which
is termed the buyer excess. The buyer excess (x) conforms to our notions
of supply and demand, yet is insensitive to the total number of speculators in the market.

The buyer excess (x) and profit (y) also provide a simple definition of the market equilibrium point. Let us consider the buyer excess
which, because there is always activity in the market, continually fluc-
tuates. It seems reasonable, however, that equilibrium would be the
point at which the number of buyers and sellers is equal; that is, when
$x = 0$. This value of the buyer excess is consistent with the random walk
model of the market which predicts that the average profit of a stock is
zero.[1,2] This consistency is a consequence of the law of supply and demand; that is, if there is no average buyer excess, there is no pressure
for the stock to change in price on the average and, therefore, the average profit is zero. The point $(0,0)$ in the (x,y)-plane is, therefore,
the point of market equilibrium.

THE MODEL

We have to construct a system of equations in terms of the buyer
excess (x) and profit (y) which will describe the dynamics of our model
market. Any finite system of equations which purports to represent a
complex physical or social situation is, at best, an approximation. Taken
literally, the equations define a model of the real system. The closer
the correspondence of the model to reality, the better the approximation
of our mathematical system to the social system. Simple models have the
advantage of being easily solved, providing immediate insight and may be
painlessly generalized. For these reasons, our model will be quite simple
while at the same time including as much of the market system as possible.

We will begin our discussion of the model by giving a general definition of what we will mean by speculator. An individual investor functions as a speculator when, in the absence of any inside information[3]
about a company, he attempts to predict the short range change in price
of that company's stock. He does this by anticipating what the average
attitude of the people in the market will be toward that particular stock.
If he predicts that people will buy soon, then the stock price will increase and he will buy now. If he forsees people selling, thereby inducing a drop in price, he will sell now, perhaps to buy later at the lower
price. The speculator's considerations are psychological and are not
based on the long term value of the company whose stock he purchases.[4]

The significant aspect of the aggregate behavior of speculators is
how their opinions and, therefore, their buying changes with time. The
simplest assumption we can make which is consistent with observed behavior is that the anticipated buyer excess at time $t+\Delta t$ is the sum of the
present buyer excess plus those that change their opinion in the interval
of time Δt.

$$x(t+\Delta t) = x(t) + \beta(t)\Delta t x(t) \qquad (2.1)$$

The units of $\beta(t)$ are inverse time and it may, therefore, be defined from
the structure of equation (2.1) to be the rate of change in the speculator's reaction to other speculators. The product $\beta(t)\Delta t$ is the fraction

of the present buyer excess who will change their opinion in the time interval Δt. Because the group of speculators is heterogeneous, in that
they each react differently to the aggregate of speculators as well as to
other changing economic conditions, the function $\beta(t)$ is assumed to be a
random function of time. The function is assumed random so as not to
bias the speculators' behavior.

It should be noted that we have shifted from the view of what an
individual speculator predicts to be the average behavior of others in
the market, to what the aggregate of speculators predict to be the aggregate behavior.[4] The average value of the function $\beta(t)$ in time, therefore, determines the average present attitude of the speculators toward
their future attitude.

If we take the continuous limit of equation (2.1); that is, for
very small time intervals Δt, we obtain the rate equation

$$\frac{dx}{dt} = \beta(t)x. \qquad (2.2)$$

If $\beta(t)$ is, on the average, positive, then the greater the buyer excess,
the greater the rate at which buyers enter the market on the average.
If, however, $\beta(t)$ is on the average negative, then the greater the buyer
excess the more rapidly buyers leave the market. (We could also have ex-
changed the equivalent term, ... sellers enter ..., for ... buyers leave
..., in the preceding sentence.) The structure of equation (2.2) results
in an exponential growth or decay of the buyer excess in time, on the
average. The random nature of $\beta(t)$ produces fluctuations about this
average behavior.

In addition to anticipating the future attitudes of other people in
the market, speculators also attempt to predict future profits of the
stock. The speculators realize that there is always a risk associated
with anticipating future attitudes of other people and so their decisions
are always tempered by the quantity of money they must invest at a given
time to realize their anticipated profit. This caution may be characterized by their reaction to the present level of profit shown by the stock.
This may be introduced into our model by modifying equation (2.1) to read

$$x(t+\Delta t) = x(t) + \beta(t)\Delta t x(t) + \alpha_1(t)\Delta t y(t) \qquad (2.3)$$

The interpretation of $\beta(t)$ is the same as in equation (2.1) but now its
effect is in competition with that of the present level of profit.

The units of $\alpha_1(t)$ in equation (2.3) are

$$\frac{\text{speculators}}{\text{profit·time}}$$

which leads to its being defined as the rate of change in the speculators'
reaction to profit. The product $\alpha_1(t)\Delta t$ is, therefore, the number of
speculators who change their buying position in the time interval Δt because of the present level of profit. Due to the variability in the effect of the level of profit on the different speculators, the function
$\alpha_1(t)$ is also assumed to be a random function of time.

If we again take the continuous limit, then equation (2.3) becomes

$$\frac{dx}{dt} = \beta(t)x + \alpha_1(t)y \qquad (2.4)$$

so that the time rate of change in the buyer excess is dependent through
the two rate functions $\alpha_1(t)$ and $\beta(t)$, on both the level of profit and

buyer excess, respectively. To determine the detailed effect of these two rate functions in equation (2.4), we must obtain the solutions to this equation. For this reason we need a second equation for the rate of growth of the profit (dy/dt).

The above analysis was psychological in content. It was concerned with how the change in size of heterogeneous group with a common goal is influenced by both size of the group and the goal level achieved by the group. Our attention must now, however, be turned to the economic mechanisms which are operative in our model market.

The first mechanism which we will assume to be operative in our model market is the law of supply and demand. If we assume for the moment that the number of shares of stock is a constant over the time interval being considered, then an increase in demand for the stock will generate an increase in the price. The form in which this will be written is to assume that a positive buyer excess induces an increase in profit. Therefore, we will write the future profit shown by a stock to be equal to the present profit plus a constant times the demand for the stock

$$y(t+\Delta t) = y(t) + \alpha_2 \Delta t x(t) \qquad (2.5)$$

The units of α_2 are

$$\frac{profit}{speculators \cdot time}$$

so that it may be defined as the rate at which the profit responds to the demand for the stock. The product $\alpha_2 \Delta t$ is a constant from one time interval Δt to another and gives the increase in profit for a given level of demand. Equation (2.5) satisfies our intuitive notions about the law of supply and demand.

In the continuous limit, equation (2.5) becomes

$$\frac{dy}{dt} = \alpha_2 x \quad . \qquad (2.6)$$

For a positive α_2 we would have an increasing rate of profit for a positive buyer excess and a decreasing rate of profit for a negative buyer excess (seller excess). It is clear, therefore, that the parameter α_2 is positive in our model.

The second mechanism assumed to be operative in our model is a random walk which was first considered in this context by Louis Bachelier in a Ph.D. thesis on the "Theory of Speculation" in 1900.[5] (More contemporary works may be found in references 1 and 2.) The form of the random walk hypothesis which we will utilize is to assume that: the change in profit in any small interval of time (Δt) is independent of the change in profit in any other small interval of time (Δt). This may be included in equation (2.5) by defining a function F(t)

$$y(t+\Delta t) = y(t) + \alpha_2 \Delta t x(t) + F(t)\Delta t \qquad (2.7)$$

where F(t) is a random function of time. The profit shown by the stock in an interval of time Δt is, therefore, the sum of its profit at the beginning of the interval, the increase generated by the buyer excess and a term which varies randomly from one interval of time Δt to another.

In the continuous limit, equation (2.7) becomes

$$\frac{dy}{dt} = \alpha_2 x + F(t) \qquad (2.8)$$

where the function of F(t) may be considered to be a random driving force which simulates the changing economic conditions of the market outside the classical notion of supply and demand. If α_2 were set equal to zero and the standard assumptions made in the Theory of Brownian Motion were assumed about F(t), then one could obtain a probability density for the profit which is Gaussian in the profit (see reference 2). We have, therefore, superimposed the law of supply and demand onto the Random Walk Model of the Stock Market.

The system of equations which define our model market is

$$\frac{dx}{dt} = \beta(t)x + \alpha_1(t)y \tag{2.9a}$$

and

$$\frac{dy}{dt} = \alpha_2 x + F(t) \tag{2.9b}$$

which is not deterministic. This model [equation (2.9)] includes the effects of market variables not considered explicitly through the random forcing term F(t). It also includes societal effects which are not necessarily economic, through the random functions $\beta(t)$ and $\alpha_1(t)$. Finally, it should be stressed that, although the law of supply and demand has been modified in equation (2.9), it maintains its deterministic form on the average.

THE FOKKER-PLANCK EQUATION

The stochastic nature of equation (2.9) implies that it is improper to describe the model market as having a precise buyer excess (x) and profit (y) at time t. Rather, we must introduce the notion of a joint probability density P(xy;t) such that

$$P(xy;t) \ dx \ dy = \text{Prob}\{x<X<x+dx, \ y<Y<y+dy;t\} \tag{3.1}$$

is the probability that the random variable X has a value between x and x+dx and the random variable Y has a value between y and y+dy at the time t. Our dynamical system must, therefore, be described in terms of the temporal evolution of the probability density P(xy;t). The equation of motion for P(xy;t) will be the Fokker-Planck equation if the random functions $\alpha_1(t)$, $\beta(t)$ and F(t) satisfy the appropriate conditions.

Consider a system which is described by the N-coordinates $x_1, x_2, \ldots, x_N$. The Fokker-Planck equation for such a system is

$$\frac{\partial P}{\partial t} = - \sum_{i=1}^{N} \frac{\partial}{\partial x_i} \{A_i P\} + \frac{1}{2} \sum_{i,j=1}^{N} \frac{\partial^2}{\partial x_i \partial x_j} \{B_{ij} P\} \tag{3.2}$$

where

$$A_i = \lim_{\Delta t \to 0} \frac{1}{\Delta t} \int\!\!\int_{-\infty}^{+\infty} (x_i - <x_i>)P(x_1, x_2, \ldots, x_N; t+\Delta t) \ dx_1 \ldots dx_N \tag{3.3}$$

$<x_i>$ is the average value of x_i taken at time t, and

$$B_{ij} = \lim_{\Delta t \to 0} \frac{1}{\Delta t} \int\!\!\int_{-\infty}^{+\infty} (x_i - <x_i>)(x_j - <x_j>)P(x_1 \ldots x_N; t+\Delta t) \ dx_1 \ldots dx_N. \tag{3.4}$$

The equation of motion, equation (3.2), describes how the probability density $P(x_1, x_2, \ldots x_N; t)$ for this system develops in time. The functions A_i and B_{ij} are the time rate of change of the mean and variance given by equations (3.3) and (3.4), respectively. The applicability of equation (3.2) to the system in question implies that correlations greater than second order in the variables $(x_1, \ldots, x_N)$ are of order $(\Delta t)^2$ and, therefore, tend to zero as $\Delta t \to 0$ faster than Δt.

The application of equation (3.2) to our model market makes a further discussion of the form of the random functions $\beta(t)$, $\alpha_1(t)$ and $F(t)$ essential. We may define an average value for each of these random functions as follows:

$$\beta(t) = \beta + \eta_1(t) \tag{3.5a}$$

$$\alpha_1(t) = \alpha_1 + \eta_2(t) \tag{3.5b}$$

and

$$F(t) = k + f(t) \tag{3.5c}$$

such that $\eta_1(t)$, $\eta_2(t)$ and $f(t)$ describe the random fluctuations about the average values β, α_1, and k, respectively. We may further assume that the average values of these fluctuations are zero; that is,

$$\langle \eta_1(t) \rangle = \langle \eta_2(t) \rangle = \langle f(t) \rangle = 0 . \tag{3.6}$$

We use equation (2.9) to determine the A_i's and B_{ij}'s for the Fokker-Planck equation (3.2) and thereby the equation of motion for the probability density of our model market. In a short time interval Δt, the average variations in the buyer excess (x) and profit (y) from equation (2.9) are:

$$\langle \Delta x \rangle = (\beta x + \alpha_1 y) \Delta t \tag{3.7a}$$

$$\langle \Delta y \rangle = \alpha_2 \, x \, \Delta t \tag{3.7b}$$

where the average value of $F(t)$ has been set equal to zero in equation (3.5c). (This may be shown to be equivalent to a change in the origin for profits which was arbitrary and may, therefore, be changed without loss in generality.) These two equations may be used to define the time rate of change in the average buyer excess and profit [equation (3.3)]:

$$A_x = \lim_{\Delta t \to 0} \frac{\langle \Delta x \rangle}{\Delta t} = \beta x + \alpha_1 y \tag{3.8a}$$

$$A_y = \lim_{\Delta t \to 0} \frac{\langle \Delta y \rangle}{\Delta t} = \alpha_2 x \quad . \tag{3.8b}$$

Also, in the same interval of time we have for the variances of our market variables,

$$\langle \Delta x \Delta x \rangle = (\beta_1 x + \alpha_1 y)^2 (\Delta t)^2 + x^2 \int_t^{t+\Delta t}\!\!\!\int \langle \eta_1(t_1) \eta_1(t_2) \rangle \, dt_1 \, dt_2$$

$$+ y^2 \int_t^{t+\Delta t}\!\!\!\int \langle \eta_2(t_1) \eta_2(t_2) \rangle dt_1 dt_2 + 2xy \int_t^{t+\Delta t}\!\!\!\int \langle \eta_1(t_1) \eta_2(t_2) \rangle dt_1 dt_2, \tag{3.9}$$

$$\langle\Delta y\Delta y\rangle = (\alpha_2 x)^2 (\Delta t)^2 + \int_t^{t+\Delta t}\!\!\int \langle F(t_1)F(t_2)\rangle\ dt_1\ dt_2 \qquad (3.10)$$

and

$$\langle\Delta x\Delta y\rangle = (\beta_1 x + \alpha_1 y)(\alpha_2 x)(\Delta t)^2 + x \int_t^{t+\Delta t}\!\!\int \langle\eta_1(t_1)F(t_2)\rangle\ dt_1\ dt_2$$

$$+\ y \int_t^{t+\Delta t}\!\!\int \langle\eta_2(t_1)F(t_2)\rangle\ dt_1\ dt_2\ . \qquad (3.11)$$

It is clear that the first terms in equations (3.9) – (3.11) all give a zero contribution to equation (3.4) since they are all of the order $(\Delta t)^2$. We must, however, examine the correlation terms in detail.

Consider a number of random functions $g_i(t)$ with zero mean ($i=1$, ...,N). We write correlations at times t_1 and t_2 as being functions only of the time difference

$$\langle g_i(t_1)g_j(t_2)\rangle = G_{ij}(t_1-t_2) = G_{ij}(\tau)\ . \qquad (3.12)$$

The correlation integrals in equations (3.9) – (3.11) can be written in the general form

$$\int_{t-t_1}^{t-t_1-\Delta t}\int_{t-t_2}^{t-t_2-\Delta t} \langle g_i(t_1)g_j(t_2)\rangle\ dt_1\ dt_2$$

$$= \int_{t-t_1}^{t-t_1-\Delta t} dt_1 \int_{t-t_2}^{t-t_2-\Delta t} G_{ij}(\tau)\ d\tau$$

$$= \Delta t \int_{t-t_2}^{t-t_2-\Delta t} G_{ij}(\tau)\ d\tau \qquad (3.13)$$

when equation (3.12) is used.

In the classical theory of Brownian Motion the function $G_{ij}(\tau)$ is set equal to a delta function in time

$$G_{ij}(\tau) = \sigma^2\delta(\tau)\delta_{ij} \qquad (3.14)$$

which would imply that any two fluctuations with a finite time separation are independent; that is, $G(\tau) = 0$, $\tau > 0$. This does not seem to be an appropriate assumption to make for $\eta_1(t)$ or $\eta_2(t)$, because an individual's memory does extend over time and, therefore, past fluctuations may influence present or future decisions as to buying behavior. It does seem appropriate, however, for the function $F(t)$ since fluctuations in economic conditions do seem to be approximately independent in time. This was the assumption made by Osborne[2] and others in their random walk theories of the stock market.

When independence of the fluctuations in time as given by equation (3.14) is not appropriate, we expand the function $G_{ij}(\tau)$ in a Taylor series

$$G_{ij}(\tau) = G_{ij}(0) + \tau\left[\frac{dG(\tau)}{d\tau}\bigg|\right]_{\tau=0} + \frac{\tau^2}{2}\left[\frac{dG(\tau)}{dt}\bigg|\right]_{\tau=0} + \ \cdots\ . \qquad (3.15)$$

The function $G_{ij}(\tau)$, therefore, is approximated in first order by the constant $G_{ij}(0)$. The higher order terms in equation (3.15) give the detailed structure of the function. Using equations (3.15) and (3.14) we obtain for the integral in equation (3.13) the two results

$$\Delta t \int_{t-t_2}^{t-t_2-\Delta t} G_{ij}(\tau) \, d\tau = \begin{cases} \sigma^2 \Delta t \delta_{ij} & (3.16a) \\ G_{ij}(0)(\Delta t)^2 + o(\Delta t^2) & (3.16b) \end{cases}$$

where equation (3.16a) corresponds to the Brownian Motion assumption and equation (3.16b) corresponds to a Taylor expansion of $G_{ij}(\tau)$ when there is a finite memory.

We may substitute equation (3.16) back into equations (3.9)-(3.11), using equation (3.16b) whenever an $n_1(t)$ or $n_2(t)$ is being correlated, and equation (3.16a) for the self-correlation of $F(t)$. This yields for the time rate of change in the variances [equation (3.4)]

$$B_{xx} = \lim_{\Delta t \to 0} \frac{<(\Delta x)^2>}{\Delta t} = 0 \qquad (3.17a)$$

$$B_{yy} = \lim_{\Delta t \to 0} \frac{<(\Delta y)^2>}{\Delta t} = \sigma^2 \qquad (3.17b)$$

$$B_{xy} = \lim_{\Delta t \to 0} \frac{<\Delta x \Delta y>}{\Delta t} = 0 \, . \qquad (3.17c)$$

The Fokker-Planck equation may now be written for our model market by substituting equation (3.8) and (3.17) in equation (3.2) and obtaining

$$\frac{\partial P}{\partial t} = -\frac{\partial}{\partial x} \{(\beta x + \alpha_1 y)P\} - \frac{\partial}{\partial y} \{\alpha_2 x P\} + \frac{\sigma^2}{2} \frac{\partial^2 P}{\partial y^2} \qquad (3.18)$$

The method of solving this equation is the topic of the next section.

THE SOLUTION

We will restrict our attention here to the class of periodic solutions which can lead to either a stable or unstable market situation. For a stable market, the average value of the random functions $\beta(t)$ and $\alpha_1(t)$ are negative. We will make use of this fact by rewriting equation (3.5) to read

$$\beta(t) = -\beta + n_1(t) \qquad (4.1a)$$

and

$$\alpha_1(t) = -\alpha_1 + n_2(t) \, . \qquad (4.1b)$$

It is then a trivial matter to change the Fokker-Planck equation to

$$\frac{\partial P}{\partial t} = -\alpha_2 x \frac{\partial P}{\partial y} + \frac{\partial}{\partial x} \{[\beta x + \alpha_1 y]P\} + \frac{\sigma^2}{2} \frac{\partial^2 P}{\partial y^2} \qquad (4.2)$$

and solve for the probability density $P(x,y;t)$ for the stable market.
To solve equation (4.2),[6] we first make the substitution of variables

$$q_1 = \alpha_2 x + \lambda_1 y \quad ; \quad q_2 = \alpha_2 x + \lambda_2 y \qquad (4.3)$$

where

$$\lambda_1 = -\beta/2 + j\omega_1 \quad ; \quad \lambda_2 = -\beta/2 - j\omega_1 \qquad (4.4)$$

and

$$\omega_1 = \sqrt{\alpha_1 \alpha_2} - \beta^2/4 \quad .$$

The parameters λ_1 and λ_2 are the roots to equation (2.9) when the random functions $\beta(t)$ and $\alpha_1(t)$ are replaced by their average values [equation (4.1)]. In terms of these new variables [equation (4.3)], the Fokker-Planck equation becomes

$$\frac{\partial P(q_1 q_2 ; t)}{\partial 2t} = - \frac{\partial}{\partial q_1} [\lambda_1 q_1 P] - \frac{\partial}{\partial q_2} [\lambda_2 q_2 P] + \frac{\sigma^2}{2} \left[\lambda_1 \frac{\partial}{\partial q_1} + \lambda_2 \frac{\partial}{\partial q_2} \right]^2 P \qquad (4.5)$$

which, because of its symmetry, is much easier to work with than equation (4.2).

To solve equation (4.5), we introduce the characteristic function $[f(\xi_1 \xi_2 ; t)]$ of the probability density $[P(q_1 q_2 ; t)]$:

$$f(\xi_1 \xi_2 ; t) = \frac{1}{2\pi} \int\int_{-\infty}^{\infty} P(q_1 q_2 ; t) \, dq_1 \, dq_2 \, \exp\{-j\xi_1 q_1 - j\xi_2 q_2\} \qquad (4.6)$$

which is the Fourier transform of the probability density. We may also write the inverse transform

$$P(q_1 q_2 ; t) = \frac{1}{2\pi} \int\int_{-\infty}^{\infty} f(\xi_1 \xi_2 ; t) \, d\xi_1 \, d\xi_2 \, \exp\{+j\xi_1 q_1 + j\xi_2 q_2\} \qquad (4.7)$$

so that a knowledge of the characteristic function determines the probability density. We may use the Fourier transform [equation (4.6)] to obtain the equation of motion for the characteristic function from equation (4.5)

$$\frac{\partial f(\xi_1 \xi_2 ; t)}{\partial t} = \lambda_1 \xi_1 \frac{\partial f}{\partial \xi_1} + \lambda_2 \xi_2 \frac{\partial f}{\partial \xi_2} - \frac{1}{2} \{\sigma_1 \xi_1 + \sigma_2 \xi_2\}^2 f \quad . \qquad (4.8)$$

where $\sigma_1 = \lambda_1 \sigma$ and $\sigma_2 = \lambda_2 \sigma$. This equation [equation (4.8)] is more easily solved than equation (4.5). The general form of the solution to equation (4.8) is

$$f(\xi_1 \xi_2 ; t) = \Phi(\xi_1 e^{\lambda_1 t}, \xi_2 e^{\lambda_2 t}) \, \exp\left\{ \frac{\sigma_1^2 \xi_1^2}{4\lambda_1} + \frac{\sigma_2^2 \xi_2^2}{4\lambda_2} + \frac{\sigma_{12}^2 \xi_1 \xi_2}{\lambda_1 + \lambda_2} \right\} \qquad (4.9)$$

where Φ is an arbitrary function of its variables and may be determined from the initial conditions.

We assume that at time $t = 0$ the variables q_1 and q_2 have the particular values q_1^0 and q_2^0. This imples that the probability density $[P(q_1 q_2 ; t)]$ is a delta function; that is,

$$P(q_1 q_2 ; t=0) = \delta(q_1 - q_1^0) \delta(q_2 - q_2^0) \quad . \qquad (4.10)$$

The characteristic function for equation (4.10), i.e., the Fourier transform, is

$$f(\xi_1 \xi_2 ; t=0) = \exp\{-j\xi_1 q_1^0 - j\xi_2 q_2^0\} \qquad (4.11)$$

so that at time $t = 0$ the function in equation (4.9) has the form

$$\Phi(\xi_1\xi_2;t=0) = \exp\{-j\xi_1q_1^0-j\xi_2q_2^0\}\exp\left\{-\frac{\sigma_1^2\xi_1^2}{4\lambda_1} - \frac{\sigma_2^2\xi_2^2}{4\lambda_2} - \frac{\sigma_{12}^2\xi_1\xi_2}{\lambda_1 + \lambda_2}\right\} \quad . \quad (4.12)$$

We may now substitute equation (4.12) with $t \neq 0$ into equation (4.9), and obtain for the chracteristic function

$$f(\xi_1\xi_2;t) = \exp\left\{\frac{\sigma_1^2(t)\xi_1^2}{2} + \frac{\sigma_1^2(t)\xi_1^2}{2} + \sigma_{12}^2(t)\xi_1\xi_2\right\}$$

$$\cdot \exp\{-j\xi_1q_1^0e^{\lambda_1 t} - j\xi_2q_2^0e^{\lambda_2 t}\} \quad (4.13)$$

where

$$\sigma_1^2(t) = \frac{\sigma_1^2}{2\lambda_1}[1 - \exp(2\lambda_1 t)]; \quad \sigma_{12}^2(t) = \frac{\sigma_1\sigma_2}{\lambda_1 + \lambda_2}[1 - \exp([\lambda_1+\lambda_2]t)];$$

$$\sigma_2^2(t) = \frac{\sigma_2^2}{2\lambda_2}[1 - \exp(2\lambda_2 t)].$$

We may therefore write the integral for the probability density as

$$P(q_1q_2;t) = \frac{1}{(2\pi)}\int\int_{-\infty}^{\infty} \exp\left\{\frac{\sigma_1^2(t)}{2}\xi_1^2 + \frac{\sigma_2^2(t)}{2}\xi_2^2 + \sigma_{12}^2(t)\xi_1\xi_2\right\} d\xi_1 \, d\xi_2$$

$$\cdot \exp\{j\xi_1\Delta q_1 + j\xi_2\Delta q_2\} \quad (4.14)$$

where

$$\Delta q_1 = q_1 - q_1^0e^{\lambda_1 t} \qquad \text{and} \qquad \Delta q_2 = q_2 - q_2^0e^{\lambda_2 t} \quad .$$

In its present form the integral in equation (4.14) appears to be divergent; however, because the real part of the roots λ_1 and λ_2 are negative, the integral will converge. The result of the integration is a bi-variate Gaussian distribution in the variables q_1 and q_2:

$$P(q_1q_2t|q_1^0q_2^0) = \frac{\exp\left\{\left[\frac{(\Delta q_1)^2}{2\sigma_1^2(t)} + \frac{(\Delta q_2)^2}{2\sigma_2^2(t)} + \frac{\rho(t)\Delta q_1\Delta q_2}{\sigma_1(t)\sigma_2(t)}\right]\Big/(1 - \rho^2(t))\right\}}{2\pi\sigma_1\sigma_2 \ \sqrt{1 - \rho^2(t)}} \quad (4.15)$$

where

$$\rho^2(t) = \sigma_{12}^4(t)/\sigma_1^2(t)\sigma_2^2(t) \quad .$$

It is also apparent that since q_1 and q_2 are linearly related to the buyer excess and profit through equation (4.3), the conditional probability density for these latter two variables is also a bi-variate Gaussian distribution.

The solution to equation (2.9) in terms of the average buyer excess and profit is now easily obtainable from equation (4.15). To see this we write the average values of q_1 and q_2

$$\langle q_1\rangle = q_1^0e^{\lambda_1 t} \quad ; \quad \langle q_2\rangle = q_2^0e^{\lambda_2 t} \quad , \quad (4.16)$$

since $\langle\Delta q_1\rangle = \langle\Delta q_2\rangle = 0$. The initial buyer excess is some value x_0, but the initial profit is zero [see equations (1.2) with $\tau = 0$]; therefore, $q_1^0 = q_2^0 = \alpha_2x_0$. Using the linear relations in equation (4.3), we find for the average buyer excess and profit

$$<x> = \frac{x_0}{\omega_1} e^{-\beta t/2} \{\omega_1 \cos\omega_1 t + \beta/2 \sin\omega_1 t\} \qquad (4.17)$$

and

$$<y> = \frac{\alpha_2 x_0}{\omega_1} e^{-\beta t/2} \sin\omega_1 t \quad . \qquad (4.18)$$

It should be noted that equations (4.17) and (4.18) satisfy the deterministic equations for our market process; that is,

$$\frac{d<x>}{dt} + \beta<x> + \alpha_1<y> = 0$$

$$\frac{d<y>}{dt} = \alpha_2<x>$$

which indicates that the mode of the bi-variate Gaussian distribution moves in a specified manner away from the initial point.

In Figures 1 and 2 the behavior of the market for both a positive and a negative value of β in equations (4.17) and (4.18) are shown,

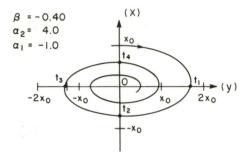

$\beta = -0.40$
$\alpha_2 = 4.0$
$\alpha_1 = -1.0$

Figure 1

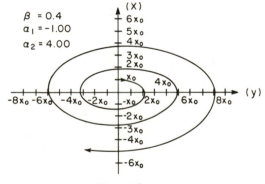

$\beta = 0.4$
$\alpha_1 = -1.00$
$\alpha_2 = 4.00$

Figure 2

respectively. The time variable has been eliminated by drawing our graph in the phase space for the system; that is, the (x,y)-plane. In both

figures the curve begins at the initial point $(x_0,0)$ at time $t = 0$ and
evolves in time, each value of x and y on the curve being at a given in-
stant of time. Therefore, the converging spiral in Figure 1 is the evolution of the market in time for $\beta > 0$ and the diverging spiral in Figure 2
for $\beta < 0$. The behavior of the market speculators which leads to the
stable market will be discussed in detail in the following section.

SUMMARY AND CONCLUSIONS

Our model was developed under the assumption that people and profit
were the important variables in a market situation. Also, that these
variables were interrelated through the Weber-Fechner law in psychology
and the law of supply and demand in economics. The people in our market
were defined to be speculators in that they invested money in stock with
a view to realizing a short term profit, regardless of the long term value
of the company in whose stocks they invested. The relative number of
buyers and sellers and the profit of a stock were used to construct a system of equations representing the dynamical behavior of our model market.
The time rate of change in the buyer excess was related to the buyer excess and the level of profit through equation (2.9a). The time rate of
change in the level of profit was given as a superposition of the law of
supply and demand and a random walk, [equation (2.9b)].

The stochastic nature of equation (2.9) required us to seek the time
dependent joint probability density $P(x,y;t)$ for our model market which
turned out to be a bi-variate Gaussian in x and y. We also found that
the average values of our market variables obeyed the same system of equations [equation (2.9)] when the random functions $\beta(t)$, $\alpha_1(t)$ and $F(t)$
were replaced by their average values.

The stable market which we discussed requires speculators to behave
in a well-defined manner. It is in specifying this behavior by the appropriate interpretation of the parameters that our model is normative. We
know from our definition of a speculator that he attempts to anticipate
what will be the future attitude of the aggregate of speculators. In a
stable market the speculator tends to believe that there will be as many
buyers for a stock as sellers, on the average. Therefore, if the magnitude of the buyer excess is other than zero, he believes that it is more
likely to decrease than increase in the immediate future. He will, therefore, sell stock if the present buyer excess is positive, anticipating a
forthcoming decrease in the buyer excess. Or, he will buy stock if the
present buyer excess is negative, anticipating a forthcoming increase in
the buyer excess. Each speculator is, therefore, trying to outguess and
act more rapidly than the other speculators and, in this way, realize a
profit. This attitude dampens the fluctuations of the buyer excess in
time and forces the market to the equilibrium point, $\langle x \rangle = 0$.

The behavior of speculators is also influenced by the profit shown
by a particular stock. Because the speculator realizes the risk he is
running in predicting what the average attitude of other speculators will
be in the future, he is reluctant to invest large sums of money when the
magnitude in profit is high. If the profit shown by a stock is high, the
speculator will be more inclined to sell than to buy since a large investment would be necessary to realize an increment of profit at this
level. Further, since he believes that the buyer excess will decrease in
the future, he would also expect the profit to decrease thus providing a
further incentive to sell. A similar argument may be presented for the
purchase of stock at negative profits.

To see how the behavior of the speculators changes in time, let us
follow the market through a cycle and pinpoint these changes. Although
the curve we will be discussing is continuous in both the average buyer

excess and profit, we will present the discussion so as to suggest a causal relation. Consider the spiral in Figure 1 which starts at time t = 0 with a positive buyer excess and zero profit [(x_0,0)].

At time $t = 0$ some speculators who are perhaps more astute than the rest realize that a positive buyer excess will eventually reverse and cause a decrease in profit and decide to sell. Because the buyer excess is large and positive, this selling generates an increase in profits. The increase in profit stimulates the more cautious speculators to sell, thereby inducing a further increase in profit. The rate at which buyers leave (sellers enter) the market continually increases in the time interval $t = 0$ to $t = t_1$. As long as the buyer excess remains positive, the profit continues to grow but at a decreasing rate. Finally, at time $t = t_1$, the profit is a maximum, there is a maximum buyer exodus from the market, the rate of growth of the profit is zero, and the number of buyers and sellers is equal.

The caution of the speculators toward both the group behavior and their reluctance to take unnecessary investment risks has retarded and eventually stopped the growth in profit in the time interval $t = 0$ to $t = t_1$. This same attitude prompts the speculators who have bought stock in this interval to begin selling at time $t = t_1$. It should be noted that, although the aggregate of speculators is assumed to have the characteristics $\beta < 0$, $\alpha_1 < 0$, the individual speculators react in a slightly different manner and at slightly different times, thus creating a continuum of activity rather than isolated bursts of action.

The continued selling at time $t = t_1$ reverses the direction of the profit and thereby stimulates more sales. The rate of profit-loss increases in the interval $t = t_1$ to $t = t_2$, whereas the rate at which sellers enter the market decreases. The rate of growth of the seller decreases because, the larger the seller excess, the more confident the speculators become that things will improve. The consequence of this is that when the maximum seller excess is reached at time $t = t_2$, it is smaller in magnitude than was the maximum buyer excess at time $t = 0$.

The profit at time $t = t_2$ is zero with respect to the $t = 0$ value, but the rate of fall in profit is a maximum. Also, the seller excess is a maximum but the rate of change in the buyer excess is zero. Those speculators who had the foresight to sell stock in the interval $t = 0$ to $t = t_1$ now re-enter the market as buyers while the profit is going negative. This action produces a braking on the fall in profit and encourages speculators to curtail sales. In the interval $t = t_2$ to $t = t_3$, the more farsighted of the speculators buy stock in ever-increasing numbers, thereby rapidly retarding the fall in profit. The sellers leave the market in increasing numbers so that at time $t = t_3$, the market has again reversed itself.

The rate at which buyers are entering the market is a maximum at time $t = t_3$ and the rate at which the profit is changing is zero. The effect of the early buying by the speculators has the effect of restricting the fall in profit to be smaller in absolute value at time $t = t_3$ to that at time $t = t_1$. The continual buying by speculators rapidly increases the rate of growth of profit just past the $t = t_3$ point.

In the interval $t = t_3$ to $t = t_4$, the rate of profit growth is increasing but the rate of growth in the buyer excess is decreasing. The negative profit continues to stimulate purchases, but the positive buyer excess makes the speculators cautious. The interference of these two effects results in the maximum buyer excess at $t = t_4$ being reduced from its $t = 0$ value. Because the market is not at equilibrium at time $t = t_4$, a second cycle of the market process is generated.

It should be stressed that in our discussion above we assumed that there were at least some of the speculators who were more clever or who knew more about the market behavior than the average speculator. This, in fact, is true. Such a person is generaly called an "insider" and has access to information which the other speculators in the market do not possess. Since I do not wish to pursue this point in detail here, I will mention only the floor specialist of the NYSE as an example of such an insider. In general an insider may be defined as any person with whom information affecting the price of the stock *originates*.

The above analysis was constrained to that of the stable market since that is generally of the most interest. It is felt that the interpretation accompanying the solutions does provide an insight into how speculators in the real market act without, of course, accounting for the myriad of details to which real speculators must attend. It is felt that these details have a tendency to cancel out over a large number of speculators and leave the average behavior of the market as discussed. The time interval over which the market returns to equilibrium, as well as the number of cycles generated, are empirical questions and cannot be answered here.

ACKNOWLEDGMENTS

The author would like to thank Professor Elliott W. Montroll for his suggestions and Professor Gerard G. Emch for reading earlier versions of this lecture and for his critical comments.

REFERENCES

1. C. W. J. Granger and O. Morgenstern, *Predictability of Stock Market Prices* (Lexington, Mass.: Heath, 1970).
2. M. F. M. Osborne, *Operations Research* 7 (1959), pp. 145-173.
3. H. G. Manne, *Insider Trading and the Stock Market* (New York: Free Press, 1966).
4. J. M. Keynes, *The General Theory of Employment, Interest and Money*, (New York: Harcourt, Brace and Company, 1936), pp. 154-156.
5. L. Bachelier, *Annals Scientifiques de l'Ecole Normale Superieure*, Sup. (3), Number 1018 (1900); English Translation by A. J. Boness in Cootner (1964), *The Random Character of the Stock Market* (MIT Press, Mass., 1964).
6. M. C. Wang and G. E. Uhlenbeck, *Rev. of Mod. Phys.* 17 (1945), pp. 323-342.

AN ENTROPY-UTILITY MODEL FOR THE SIZE DISTRIBUTION OF INCOME

Wade W. Badger

Institute for Fundamental Studies
University of Rochester

INTRODUCTION

The size distribution of incomes has fascinated economist, sociologists and statisticians for a long time. To both develop the background and present a new approach to the problem cannot be done in a single lecture. I have, therefore, divided my talk into two lectures. The first will selectively review previous theoretical and empirical work on the size distribution of incomes as well as introduce the new methodology. The second lecture will present the theoretical form derived for the size distribution of incomes by means of this methodology and discuss its statistical fit to income data (specifically 1935-36 U.S. data) at some length. The basis of my approach to this problem has its origins in statistical mechanics, which is not really surprising because the author's formal training has been in physics rather than economics. It should be added that the conceptual foundation involved has been given formal independent standing by information theorists, although I will present only those essentials of the formalism which are relevant to the problem at hand.

ECONOMIC BACKGROUND

Probably because economists first focused their attention on the relative income shares accruing to the factors of production (mainly capital and labor), it was not until the 1890's that Vilfredo Pareto made the first extensive statistical study of the size distribution of incomes. The empirical law which Pareto postulated as a result of his studies is usually given as

$$N = Ax^{-\nu} \tag{1}$$

where A is a proportionality constant, ν is the Pareto coefficient, and N is the number of people having an income x or larger. We will be stretching standard mathematical practice only slightly if we rewrite equation (1) as

$$1 - P(x) = A'x^{-\nu} \tag{2}$$

Here A' is A divided by the total number of income recipients and P(x)

may be considered the expectation value of the proportion of these people
having an income which is less than or equal to x.

Pareto found that a single value of the parameter ν(=1.15) very
nearly characterized every society and every period for which data were
available. Various values of ν, as determined by Pareto (Davis, p. 30)
are given in Table 1. The income distribution p(x) = dP(x)/dx (the

TABLE 1

PARETO COEFFICIENTS FOR VARIOUS PLACES AND DATES

Country		ν	Country		ν
England	(1843)	1.50	Perugia (city)		1.69
	(1879–80)	1.35	Perugia (country)		1.37
	(1893–94)	1.50	Ancona,Arezzo,Parma,Pisa		1.32
Prussia	(1852)	1.89	Italian cities		1.45
	(1876)	1.72	Basel		1.24
	(1881)	1.73	Paris (rents)		1.57
	(1886)	1.68	Augsburg	(1471)	1.43
	(1890)	1.60		(1498)	1.47
	(1894)	1.60		(1512)	1.26
Saxony	(1880)	1.58		(1526)	1.13
	(1886)	1.51	Peru (at the end of		1.79
Florence		1.41	18th century)		

expectation value of the proportion of people having incomes between x and
x + dx) implied by equation (2) is monotonically decreasing with increasing x. Given the high income data which Pareto had available, this would
appear quite reasonable. However, more recent data show that income distributions typically have a single mode at a relatively low value of x.
Pareto's law is the only valid empirical description of the high income
tail of the distribution, i.e., those incomes above some ill-defined
threshold value. In practice, the Pareto distribution is most often used
to graduate the incomes of the top 5% of all consumer units and becomes
quite inaccurate if used to graduate more than the top thirty or forty
percent of such units.

The following U.S. data for 1935–1936 (Table 2) will hopefully give
the reader some feeling for the type of distribution under consideration
(U.S. Nat. Resources Committee, p. 6). A graph of the data in Table 2 is
not presented here because presupposing an arithmetic scale of 1/8th inch
= $1,000, it would have to extend over ten feet just to include the
$500,000 to $1,000,000 category. This also emphasizes the *extreme* skewness of the distribution. Utilizing even the most skewed biometric distributions (those associated with Karl Pearson, to be more precise), one
rarely finds an event which is more than four standard deviations from the
mean. In the case of personal incomes, however, individuals are regularly
found whose incomes are several hundred standard deviations from the mean.

It is often said (Klein, p. 156; Quensel) that the lognormal distribution given in equation (3) fits income data better than the Pareto distribution at the low end of the income scale, while the reverse is true
at the upper end of the income scale

$$P_{LN}(x) \, dx = \frac{1}{\sqrt{2\pi} \; \sigma_{LN}} \, e^{-(1/2\sigma_{LN}^2)(\ln x - \mu_{LN})^2} \, \frac{dx}{x} \qquad (3)$$

The lognormal distribution is characterized by a centering parameter

TABLE 2

DISTRIBUTION OF FAMILIES AND SINGLE INDIVIDUALS AND OF AGGREGATE
INCOME RECEIVED, BY INCOME LEVEL, 1935-1936

Income Level	Families and single individuals		Aggregate Income	
	Number	Cumulative%	Amount (in thousands)	Cumulative
Under $250	2,123,534	5.38	$249,138	0.05
$250-500	4,587,377	17.01	1,767,138	3.48
$500-750	5,771,960	31.64	3,615,653	9.58
$750-1,000	5,876,078	46.53	5,129,506	18.23
$1,000-1,250	4,990,995	59.18	5,589,111	27.65
$1,250-1,500	3,743,428	68.66	5,109,112	36.27
$1,500-1,750	2,889,904	75.99	1,660,793	44.14
$1,750-2,000	2,296,022	81.81	4,214,203	51.25
$2,000-2,250	1,704,535	86.13	3,602,861	57.33
$2,250-2,500	1,254,076	89.30	2,968,932	62.34
$2,500-3,000	1,475,474	93.04	4,004,774	69.10
$3,000-3,500	851,919	95.20	2,735,487	73.72
$3,500-4,000	502,159	96.48	1,863,384	76.86
$4,000-4,500	286,053	97.20	1,202,826	78.89
$4,500-5,000	178,138	97.65	811,766	80.31
$5,000-7,500	380,266	98.62	2,244,406	84.10
$7,500-10,000	215,642	99.16	1,847,820	87.22
$10,000-15,000	152,682	99.55	1,746,925	90.17
$15,000-20,000	67,923	99.72	1,174,574	92.15
$20,000-25,000	39,825	99.82	889,114	93.65
$25,000-30,000	25,583	99.89	720,268	94.87
$30,000-40,000	17,959	99.93	641,272	95.95
$40,000-50,000	8,340	99.95	390,311	96.61
$50,000-100,000	13,041	99.986	908,485	98.14
$100,000-250,000	4,144	99.997	539,006	99.05

$$\mu_{LN} = \overline{\log x} = \int_0^\infty \log x \ P_{LN}(x) \ dx \tag{4}$$

which can also be interpreted as the log of the geometric mean income, and
a scale parameter σ_{LN} satisfying the relation

$$\sigma_{LN}^2 = \int_0^\infty (\log x - \overline{\log x})^2 \ P_{LN}(x) \ dx \quad . \tag{5}$$

As we shall soon see, σ_{LN} has been closely associated with the concentration or inequality of incomes characterizing the probability density function P(x). Careful examination of U.S. data shows that the lognormal distribution fits these data quite well from approximately the twentieth percentile to the ninety-fifth percentile (Kravis, p. 172). The bottom 20%
of all income recipients are receiving less than would be expected according to the lognormal distribution, whereas the top 5% are receiving relatively more than would be expected.

The lognormal distribution also distinguishes itself from the Pareto distribution in that it can be derived from the "law of proportionate effect." This law states that the change in the variate (in this case, an individual's income) at any stage is a random proportion of its present value. This can be expressed as

$$x_j - x_{j-1} = \varepsilon_j x_{j-1}$$

where the set of random variables $\{\varepsilon_j\}$ are mutually independent and, furthermore, independent of all the x_j's. After a sequence of n proportionate "random shocks," an individual's income will be

$$x_n = x_0(1 + \varepsilon_1)(1 + \varepsilon_2)...(1 + \varepsilon_n)$$

where x_0 was his initial income at some arbitrary time origin. If we take the log of x_n, we obtain

$$\log x_n = \log x_0 + \log(1 + \varepsilon_1) + \log(1 + \varepsilon_2)... + \log(1 + \varepsilon_n) \ .$$

Because each term on the right hand side of this equation is an independent random variable, one can deduce that $\log x_n$ will be normally distributed for sufficiently large n by invoking the appropriate form of the Central Limit Theorem (Aitchison and Brown, pp. 22–23).

Economists, especially welfare economists, have long sought a scalar measure which would appropriately characterize the vast income inequalities between the rich and the poor. Pareto's law aroused so much controversy because of the extreme rigidity it implied about the relative distribution of income. Pareto took ν as the sought-after measure of inequality. According to Pareto, inequality increased as ν increased, although succeeding economists have usually taken the opposite point of view. Assuming that ν remains constant, it follows directly that, "Neither an increase in the minimum income nor a diminution in the inequality can come about, except when the total income increases more rapidly than the population" (Dalton, Appendix, p. 13). However, partly because none of the distributions proposed so far are "obviously" superior when it comes to fitting income data, no single statistic or distributional parameter has gained general acceptance as *the* measure of inequality to the exclusion of others.

The Pareto coefficient ν and the quantile measures (e.g., the percentage of the total income earned by the top or bottom 5% of all income recipients) are sometimes used for convenience and/or historical continuity, but are insensitive to the structure of one or more parts of the distribution. The first absolute moment about the mean

$$\beta_1 = \sum_{i=1}^{n} | x_i - \bar{x}_i | p(x_i)$$

is only infrequently used as a measure of inequality and fails to meet the condition imposed by Dalton's "law of transfers." This law stipulates that inequality (and any measure thereof) should decrease whenever there is a transfer of income from a richer individual to a poorer individual which does not reverse or more than reverse their relative positions (Dalton, Appendix, p. 5). As a measure of inequality, β_1 is completely insensitive to transfers (i.e., it remains constant) as long as the two incomes in question are both above or both below the mean $\bar{x}_i$. The standard deviation σ, where

$$\sigma^2 = \sum_{i=1}^{n} (x_i - \bar{x}_i)^2 p(x_i) \tag{6}$$

and

$$\bar{x}_i = \sum_{i=1}^{n} x_i p(x_i) \tag{7}$$

is, however, perfectly sensitive to transfers and is often used as a measure of inequality. Alternately, the coefficient of variation $\sigma\sqrt{x_i}$ can be utilized if one wishes to define a measure which is independent of the unit in which income is measured (Yntema, pp. 423-425). A measure based on deviations from the mean instead of some other central value is justified on the grounds that this value would obtain if complete equality were reached. No use is made in the literature of third or higher moments about the mean as measures of inequality, presumably because of the extreme skewness characteristic of income distributions and the resulting sensitivity which such measures would have to income changes in the highest brackets.

The standard deviation of the logarithm of income (σ log), where

$$(\sigma \ \text{log})^2 = \sum_{i=1}^{n} (\text{log } x_i - \overline{\text{log } x_i})^2 p(x_i) \tag{8}$$

and

$$\overline{\text{log } x_i} = \sum_{i=1}^{n} (\text{log } x_i) p(x_i) \tag{9}$$

has been extensively used as a measure of inequality. In the first place, it reduces the relative importance of the high income brackets and is analogous to the coefficient of variation in being invariant to the unit in which income is measured. Secondly, the logarithm of income has a welfare significance which stems originally from D. Bernoulli's (1738) famous postulate relating income to the satisfaction which it engenders. This postulate assumed that

$$dU = k \ dx/x \tag{10a}$$

where dU is the additional satisfaction (or utility or social well-being) derived from an additional income increment dx given an initial income x. The proportionality constant k is assumed to vary from individual to individual. An individual's total utility U(x) would then be

$$U(x) = k \ \text{log } x/x_0 \tag{10b}$$

where x_0 is some threshold value which is necessary to just sustain life. Thus the use of σ log as a measure of inequality is sometimes jusitifed because "Equal percentage differences in income [i.e., equal increments of log x] at two different points on the income scale are apt, we believe, to have more nearly the same welfare significance than equal absolute differences" (Kravis, p. 179).

There exists a third reason for using σ log as a measure of inequality which has to do with its connection to the lognormal distribution. If the general probability density function $p(x_i)$ in equation (6) is taken to be the lognormal distribution, then σ log is identically equal to the dispersion parameter σ_{LN} given in equation (5). As the derivation of the lognormal distribution makes use of sequential "random shocks," one must

make some additional assumptions (Kalecki, p. 162) about the type of process taking place. Without them σ_{LN} would, in theory, increase monotonically with time (i.e., with the number of random shocks). In point of
fact, calculations utilizing post-war U.S. data indicate that σ log has
slowly oscillated about a central value over this entire period (Schultz,
p. 79).

Perhaps the most frequently used measure of inequality is that due
to Lorentz. He made use of a particular graphical technique which related the proportion of income recipients having income less than x (measured along the horizontal axis) to the proportion of total income accruing
to these same income recipients (measured along the vertical axis). The
Lorentz (or Gini) measure is then the ratio of the area between the curve
so traced out and the "diagonal of equal distribution" to the total area
of the right triangle under this same diagonal. It is interesting that
there exists a simple one-to-one relation between the Lorentz measure L
and the variance parameter σ_{LN} in the lognormal distribution, namely

$$L = 2N \left(\frac{\sigma_{LN}}{\sqrt{2}} \,\middle|\, 0,1 \right) - 1$$

where $N(x|\mu,\sigma^2)$ is the normal distribution function having mean μ and
variance σ^2 (Aitchison and Brown, p. 112). It should be noted, however,
that L and σ log are theoretically independent measures of inequality and
become effectively equivalent only when or to the extent that income data
are lognormally distributed.

If we attempt to estimate income inequalities or the temporal changes
in inequality using any of the measures discussed above we immediately
run into problems of comparability. Our measures of inequality vary significantly with the definition of income (e.g., should the market value
of non-money income be included?), the classification of income-receiving
units (e.g., should said units be individuals, families, spending units,
or what?), and the technique used in gathering data (e.g., surveys give
more accurate statistics for lower income groups while income tax data
provide a more accurate picture of the higher income groups). In fact,
for any of the measures of inequality which have earned some measure of
general acceptance and which have no major theoretical drawbacks (namely,
L, σ, and σ log), variations due to the above definitional problems "are
nearly as wide as those which are found between countries near the extremes of equality and inequality..." (Kravis, p. 182). There is also
some evidence that the relative rankings assigned to a sequence of income
distributions will vary significantly depending on the measure of inequality used (Yntema, pp. 429-431). Insofar as valid intertemporal comparisons can be made using a single measure no lasting or impressive changes
appear to have taken place. The relatively large decrease in U.S. income
inequality during the period 1935-36 to 1944 should be noted as a possible
exception. Such a decrease would be quite understandable in light of New
Deal policies and war pressures which combined to greatly increase income
taxes and the level of transfers (Cartter). The accuracy of the above
generalizations is, of course, open to debate, but all in all one gets
the impression that income inequalities have remained remarkably constant.
This is perhaps all the more interesting because

> "The drift of economic science during many generations has been
> with increasing force towards the belief that there is no real
> necessity, and therefore no moral justification, for extreme
> poverty side by side with great wealth." (Marshall)

In short, the long standing pressure toward egalitarianism has not produced much in the way of results. On the other hand, the ever-increasing inequality which Marxists predicted would lead to class revolution has also failed to materialize.

THE ENTROPY APPROACH

Given the foregoing brief review of the problem, I would like to introduce a general methodology for modeling complex systems, which I shall term the entropy approach. Suppose we are interested in the relative distribution of some important attribute of a system X which, considered as a random variable, can take on a finite number of discrete values $x_1, x_2, x_3 \ldots x_n$ with probabilities $p(x_1), p(x_2), p(x_3), \ldots p(x_n)$. The discrete values or states are delineated as part of the problem, but the corresponding probabilities are unknown. A certain amount of information is available to us about the system, namely, a finite number of constraining relations of the form

$$\sum_i p(x_i)[f_j(x_i)] = c_j \qquad (j = 1, 2, \ldots, m) \qquad (11)$$

where the function $f_j(x_i)$ and constants c_j are given for each of m-independent relations. In order for our probability interpretation to remain consistent, we must require one of the constraints to be of the form

$$\sum_i p(x_i) = 1 \; . \qquad (12)$$

Under these circumstances, what can we say about the set of probabilities $\{p(x_i)\}$? Probabilists of the "objective" school of thought would refrain from saying anything about the $p(x_i)$ unless the number of constraining equations in (11) was one less than the number of states. Faced with situations in which $n \gg m+1$, information theory can at least provide us with a subjective estimate of the set $\{p(x_i)\}$ through the introduction of entropy.

The entropy measure introduced by Shannon

$$S(p_1, p_2, \ldots, p_n) = - k \sum_i p(x_i) \, \ell n \, p(x_1) \qquad (13)$$

gives us an unique, unambiguous measure of the amount of uncertainty connected with a discrete probability distribution. Further, it provides us with the "best" (i.e., the most probable and least biased) estimate we can make for $\{p(x_i)\}$ on the basis of the information available, i.e., that which maximizes equation (13) subject to the constraints given in equations (11) and (12) (Jaynes, 1957, pp. 621-623). This maximization procedure straightforwardly yields

$$p(x_i) = e^{-\lambda_0 - \sum_{j=1}^{m} \lambda_j f_j(x_i)} \qquad (14)$$

where λ_0 is the Lagrange multiplier associated with and determined by equation (12) and the λ_j's are the comparable multipliers associated with the relations given in equation (11). Because equation (12) is just the normalization condition, it follows immediately that

$$e^{\lambda_0} = \sum_i e^{-\sum_j \lambda_j f_j(x_i)}$$

(15)

The functions $f_j(x_i)$ and the numbers c_j in equation (11) should be interpreted as constant in time. It therefore follows that the entropy approach yields an equilibrium distribution for $\{p(x_i)\}$ that is, one that will hold for all time so long as the system is not disturbed. The theoretical question of how to deal with a system that is continuously evolving in time and/or one which is affected by external influences has not, in general, been successfully answered. Several approaches to this problem exist and I will in due course discuss one which appears useful in the economic context at hand.

Constrained entropy maximization has already been found useful in such diverse fields as statistical mechanics (Jaynes, 1957) and urban system's modeling (Wilson, 1969, 1970). Jaynes' (1957) reformulation of equilibrium statistical mechanics uses constrained entropy maximization as a starting point, whereas the procedure usually followed brings entropy into the picture at the last step via a comparison with phenomenological thermodynamics. The most basic treatment assumes one constraining equation namely, one specifying that the expected value of the energy of a closed system is constant. The associated Lagrange multiplier can then be identified as $1/kT$ where k is the Botlzmann constant and T is the temperature of the system. In the case of the quantum mechanical grand canonical ensemble, specification of a possible state requires that the number of several types of molecules be known in addition to the energy of the system. The expected values of these numbers then enter as additional constraining equations and the associated Lagrange multipliers turn out to be the chemical potentials.

The entropy approach implicitly assumes the presence of an underlying stochastic process and/or conceptually necessitates dealing with an unwieldy deterministic system in a probabilistic manner. Its great power and, at the same time, its pitfall as a model building technique is its ability to provide the "most probable" answers (i.e., distribution functions) without having to solve the frequently intractible equations generated otherwise. The danger lies in the possibility of an inappropriate or insufficient specification of the constraining equations and the possibility of an inappropriate or incomplete enumeration of all possible states. Let us first deal with the constraints. Jaynes (1957, p. 623) writes that "The maximum entropy distribution may be asserted for the positive reason that it is uniquely determined as the one which is maximally noncommittal with regard to missing information, instead of the negative one that there is no reason to think otherwise." We could perhaps, be even more positive and accept the maximum entropy distribution to the extent that we believe the constraining equations reflect the fundamental nature of the system. We might even make this assertion while possessing further knowledge about the system which was, however, believed peripheral to the basic mechanisms at work. Unfortunately, for all too many complex systems, there exists no definitive basis for distinguishing between what is basic and what is peripheral. Looking at this problem in a positive light, however, this juncture provides an important degree of modeling freedom whereby the consequences of adding constraints, i.e., additional knowledge, can be made immediately apparent through the changes produced in the corresponding maximum entropy distributions. One can, of course, only expect significant agreement with phenomenological distributions to the extent that there exists a sharp differentiation between a small number of fundamental constraints and everything else which (to the

extent that our methodology mirrors reality) behaves in nearly as random
a fashion as possible.

Proper enumeration of the possible states is as important as, if not
more important than, appropriately specifying the constraints. Note first
that if the only constraint specified is the normalization condition (12),
equation (15) tells us that

$$p(x_i) = e^{-\lambda_0} = 1/n \qquad (16)$$

where n is the total number of states. Each of the states is then *a
priori* equally probable. Note that if equation (16) obtains, then S =
k log n and the entropy or uncertainty is monotonically increasing with
n and vanishes (i.e., there is no uncertainty) when n = 1. Our uncertainty about a system increases with the number of *a priori* equally probable states which the system may be in. Relation (16) is in accordance
with our intuitive notion that a broad probability distribution (which in
the extreme case means a "flat" distribution) represents more uncertainty
than a sharply peaked one (which in the extreme case is a single "spike").
States must then be chosen so that they are in some basic sense equally
important. Such a choice calls for a deep fundamental understanding of
the particular problem at hand. The great initial advances that took
place in physics with the introduction of quantum theory can quite cor-
rectly be looked at as a clarification of the nature of the state space,
rather than any fundamental change in the constraining equations, e.g.,
the conservation of energy.

AN ENTROPY-UTILITY MODEL FOR THE SIZE DISTRIBUTION OF INCOMES

Specifying the state space turns out to be the most difficult part
of utilizing constrained entropy maximization to rationalize the size
distribution of incomes. We appear to be presented with a continuous
range of possibilities extending from some quite small (and operationally
quite difficult to determine) minimum value of income (x_{min}) to the total
income x_{total} earned by everyone in the country (or other unit we might
be considering). In order to obtain a finite number of discrete states
to which our theoretical treatment can then be applied, let us divide up
the intervale x_{min} to x_{total} into a conveniently large but finite number
of segments n and use the values characterizing the midpoints of successive segments as discrete states. Such a procedure is suggested by the
fact that personal income data are usually aggregated into classes or income "brackets" for ease of presentation. There are in infinite number
of ways in which we can space these divisions and we can quite generally
represent them as equal increments of some monotonically increasing function y = y(x). The trick is to choose our "independent variable" y(x) in
such a way that equal increments of y(x), i.e., $\Delta y(x) = \varepsilon$, will have equal
importance, i.e., equal *a priori* weightings.

At this point I shall make the *Ansatz* that

$$y(x) \equiv U(x) \qquad (17)$$

where U(x) is a utility function of the type suggested by von Neumann
and Morgenstern (1947) and further developed by Friedman and Savage
(1948). As a normative theory of economic behavior, utility maximization
assumes that, under conditions involving uncertainty, individuals will
choose the jth course of action or "prospect" $\{p^j(x_1), p^j(x_2), \ldots, p^j(x_n)\}$
as if they were maximizing the expected value of a utility function $U(x_i)$

$$\max_{j}\{\overline{U^j(x)}\} = \max_{j}\{\sum_{i} p^j(x_i)U(x_i)\} \qquad (18)$$

where $p^j(x_i)$ is the probability of attaining income x_i having chosen prospect j. Utility maximization can be considered a generalization of profit
maximization (or maximization of the expected value)

$$\max_{j}\{\overline{x^i}\} = \max_{j}\{\sum_{i} p^j(x_i)x_i\} \qquad (19)$$

which takes account of an individual's attitude toward risk and [provided
$U(x)$ is unequal to x or $e^{\alpha x}$] his relative income position (Borch, p. 45).
In fact, relation (18) quite naturally reduces to (19) if these two factors are discounted and we put our trust solely in the Law of Large
Numbers.

The concept of utility (which has at times been interpreted as satisfaction or value) has had a turbulent economic career. It was first
suggested by an eighteenth century mathematician named Craemer (Bernoulli,
p. 33). His contemporary, D. Bernoulli, developed the idea at some length
and introduced the specific form of $U(x)$ given in (10b) in order to explain the reluctance of most people to pay more than a small amount for
the opportunity to play a game of chance with an infinite expected value
(The St. Petersburg Paradox). Utility maximization subsequently became
intimately connected with a strong belief in diminishing marginal utilities, a belief that $d^2U/dx^2 < 0$ for all x; certainly a dollar meant more
to a rich man than a poor man. Unfortunately this pair of concepts could
not, in general, consistently explain the type of gambling behavior which
originally motivated the introduction of (10b). Furthermore, a completely
satisfactory theory of riskless choice was built up using only the ordinal
properties of utility. The concept of utility maximization thus fell into
general disrepute until, stripped of the necessity of diminishing marginal
utilities, von Neumann and Morgenstern (1947) gave it a firm axiomatic
basis. This set of postulates provides simple, intuitively appealing arguments as to why the rational economic man should indeed act as if he
were maximizing the expected value (Bernoulli would have said "moral expectation") of something called an utility function. Such a function as-
sociates finite real numbers $U(x_i)$ with each of the possible outcomes,
x_i, but can only be defined up to a positive linear transformation

$$U'(x_i) = k_1 U(x_i) + k_2 \qquad (20)$$

where k_1 and k_2 are real constants, $k_1 > 0$. The most that can be said
about an individual's behavior is just as well represented by $U(x_i)$ as
any $U'(x_i)$. Von Neumann and Morgenstern assumed that the probabilities
$p(x_i)$ could be objectively defined, whereas subsequent normative work
[which attains what is perhaps its most refined statement in the work of
Leonard Savage (1954)] has stressed individually determined subjective
probabilities.

Thus the concept of utility maximization implies that $U(x)$ is the
natural or linearized scale in terms of which economic value judgments
are made [see equation (18)]. It is then, I think quite reasonable to
associate equal *a priori* importance to equal intervals of $U(x)$ and thus
to accept equation (17). Unfortunately for this point of view, modern
utility theory is not based on the existance of a single utility function
$U(x)$, but rather admits the possibility of different utility functions
for each and every individual. In order to make sense out of our entropy-
utility approach, we must face the difficulty of finding a *group mean
utility function* $U(x)$ given a set of individual utility functions $\{U_I(x)\}$.

It is evident that individual members of $\{U_I(x)\}$ should have reasonably similar shapes just because the individuals in question are interacting in the same economic milieu. Indeed this is the attitude taken by Friedman and Savage (1948, p. 299). I have found it meaningful to represent the averaging process in question by the relation

$$U(x,a_1,a_2,\ldots,a_n) = \int U_I(x,a_1',a_2',\ldots,a_n') \, dF(a_1',a_2',\ldots,a_n') \qquad (21)$$

where $F(a_1',a_2',\ldots,a_n')$ is a joint distribution function over a set of random variables $\{a_i'\}$ which parametrically determine the explicit form of the individual utility functions. Given a world of complete information and perfect markets populated by purely economic beings possessing unlimited mental capacities, there is every reason to believe that $F(a_1',a_2',\ldots,a_n')$ would reduce to an n-dimensional delta function

$$F(a_1',a_2',\ldots,a_n') \to \delta(a_1'-a_1)\delta(a_2'-a_2)\ldots\delta(a_n'-a_n)$$

which can be considered a kind of supra-rationality condition. The set of individual utility functions would then reduce identically to $U(x,a_1, a_2,\ldots,a_n)$ from which it is seen that the functional form of each member of $\{U_I(x,a_1',a_2',\ldots,a_n')\}$ may be considered identical with that of the group mean utility function.

Time limitations prohibit me from developing the implications of equation (21) which, in due course, implicitly suggest a particular form for $U(x)$. The following sections will introduce what I believe to be the appropriate functional form and accompanying parameterization of $U(x)$. I would like to point out, however, that the relative weightings given by $F(a_1',a_2',\ldots,a_n')$ are not to be construed as direct trade-offs between individual interests or the "interpersonal comparisons of utility" which have plagued game theorists (Luce and Raiffa, p. 34). The proper interpretation of these weightings is thought to be that they represent the relative frequency with which individuals make decisions in accordance with a given attitude toward risk. Of course, the very existence of "the market" implies collective decision making and concomitant trade-offs between individual interests. It is hoped, however, that $U(x)$ can be considered meaningful in terms of the microeconomic limit, as indicative of how the "average" man will behave under exogeneously specified market conditions.

We are fortunate that previous work on the size distribution of income makes the appropriate macroeconimic constraints somewhat easier to specify than the appropriate state space. Our previous discussion of parameters which measure income inequality quite naturally suggests an inequality constraint of the form

$$\sum_{i=1}^{n} [U(x_i) - \overline{U(x_i)}]^2 p[U(x_i)] = \text{constant} = c_1 \qquad (22)$$

where

$$\overline{U(x_i)} = \sum_{i=1}^{n} U(x_i)p[U(x_i)]. \qquad (23)$$

The states $i = 1,2,\ldots,n$ are formed by dividing the interval $[U(x_{min}), U(x_{total})]$ into n equal increments of length $\Delta U = \varepsilon$ and letting $U(x_i)$ equal half the sum of the values of $U(x)$ at the end points of the respective (ith) interval. Note that σ and $\sigma \log$ are of the general form given in equation (22), if we allow for the fact that the income brackets or

states are often determined arbitrarily rather than by the method specified above. As mentioned previously, the Lorentz measure is also connected to a specific form of (22) through the lognormal distribution.

If equation (22) was our only constraint, then equation (14) would
yield the following maximum entropy distribution

$$p[U(x_i)] = \exp\{-\lambda_0 - \lambda_1[U(x_i) - \overline{U(x_i)}]^2\} \qquad (24)$$

If we make use of equation (15) and let $\lambda_1 \equiv 1/2\sigma_U^2$, equation (24)
becomes

$$p[U(x_i)] = \frac{\exp\{-[U(x_i)-\overline{U(x_i)}]^2/\sigma_U^2\}}{\sum\limits_{i=1}^{m} \exp\{-[U(x_i)-\overline{U(x_i)}]^2/2\sigma_U^2\}} . \qquad (25)$$

The entropy formalism has a rigorous basis only if the number of states,
or possible incomes, is finite. However, at this point it does no harm
to pass to the continuous limit and obtain

$$p[U(x_i)] \rightarrow p[U(x)] \; dU = \frac{\exp\{-[U(x)-\overline{U(x)}]^2/2\sigma_U^2\} \; dU}{\int\limits_{U(x_{min})}^{U(x_{total})} \exp\{-[U(x)-\overline{U(x)}]^2/2\sigma_U^2\} \; dU} \qquad (26)$$

where

$$\overline{U(x)} = \int\limits_{U(x_{min})}^{U(x_{total})} U(x)p[U(x)] \; dU, \quad [U(x_{min}) \leq U(x) < U(x_{total})]. \qquad (27)$$

Writing this probability density function in terms of income x instead of
utility $U(x)$ introduces a density of states factor $\rho = dU/dx$ which is a
direct result of having chosen our state space in accordance with equation (17). Thus

$$p(x) \; dx = p[U(x)] \; dU = \frac{\exp\{-[U(x)-\overline{U(x)}]^2/2\sigma_U^2\} \; \rho(x) \; dx}{\int\limits_{x_{min}}^{x_{total}} \exp\{-[U(x)-\overline{U(x)}]^2/2\sigma_U^2 \; \rho(x) \; dx} \qquad (28)$$

where

$$\overline{U(x)} = \int\limits_{x_{min}}^{x_{total}} U(x)p(x) \; dx, \quad (x_{min} \leq x \leq x_{total}) . \qquad (29)$$

Most often the presentation of income data makes the choice of $x_{min} = 0$
quite natural. Alternately, we can take x_{min} as our origin of measurement and rewrite our equations in terms of $x' = x - x_{min}$. The empirical
values of $_U$ and $\overline{U}$ may also be of such a magnitude (as in the case of U
$= \log x$) that the amount of probability weight above x_{total} is absolutely
negligible and the difference between upper limits of x_{total} and infinity
in the denominator of (28) will therefore be inconsequential. Under these
circumstances, (28) and (29) become

$$p(x') \; dx' = \frac{\exp\{-[U(x') - \overline{U(x')}]^2/2\sigma_U^2\} \; \rho(x') \; dx'}{\int\limits_{0}^{\infty} \exp\{-[U(x') - \overline{U(x')}]^2/2\sigma_U^2\} \; \rho(x') \; dx'} \qquad (30)$$

and

$$\overline{U(x')} = \int_0^\infty U(x')p(x')\,dx' \quad , \quad 0 \le x' = x - x_{min} \tag{31}$$

Note that equation (30) reduces to a truncated normal distribution if we go along with the classical assumption of profit maximization [see equation (19)]. Bernoulli's hypothesis [see equation (10b)] recommends itself to us as one of the few functional forms (the others being functions of log x') which completely avoids the truncation problem. More importantly, equation (28) then reduces to the lognormal distribution

$$p(x')dx' = \frac{\exp\{-(\log x' - \mu_{LN})^2/2\sigma_{LN}^2\}}{\sigma_{LN}(2\pi)^{\frac{1}{2}}}\ \frac{dx'}{x'} \quad ; \ 0 \le x' = x - x_{min} \le \infty$$

or its three-parameter generalization (Aitchison and Brown, p. 14) if $x_{min} \ne 0$.

A HEURISTIC DETERMINATION OF U(x)

I would like to indicate how one may arrive at the appropriate functional form of U(x) starting from some older work on riskless consumer choice. This work attempted to measure marginal utilities dU/dx over selected income ranges, given the validity of certain independence and uniqueness assumptions (Vickrey, pp. 319-324). Consequently, the results obtained in this manner are not necessarily compatible with von Neumann and Morgenstern's "risky choice" approach. Nevertheless, these results do give us an analytical starting point in our search for U(x) which, because of the current stress on individual decision-making is not generally present in the more recent literature. Ragnar Frisch (1932, p. 31) has summarized and slightly reinterpreted this work by considering the following general form

$$\frac{dU}{dx} = kG(x, x_{min}) \tag{32}$$

for the marginal utility. Bernoulli's contribution [see equation (10a)] was reinterpreted as

$$G(x, x_{min}) = \frac{1}{(x - x_{min})} \quad\quad \text{(Bernoulli)}$$

whereas Dalton (Appendix, p. 4) and, apparently independently, Jordan (1924, p. 189) proposed a form which is basically

$$G(x, x_{min}) = \frac{1}{(x - x_{min})^2} \quad\quad \text{(Dalton, Jordan)}$$

Frisch (1932, p. 31) himself preferred

$$G(x, x_{min}) = \frac{1}{(\log x - \log x_{min})} \quad\quad \text{(Frisch)} \quad .$$

The major point of conflict between the modern viewpoint of utility theory and the specific forms of U(x) summarized by Frisch, is the inadmissibility of infinite values of U(x) in the former context and the presence of such values in the latter. In fact, Menger (1934) concluded that U(x) must necessarily be bounded even before the publication of *The Theory of Games and Economic Behavior*. Note that dU/dx becomes infinite at x = x_{min} and consequently $U(x_{min}) = -\infty$ for all the forms of $G(x,x_{min})$ proposed above. In order for a utility function to take on an infinite value at one or more points, one or more of the fundamental postulates upon which modern utility theory is based would have to break down. To the extent that we can identify and operationally validate these same postulates we can indeed establish the inadmissibility of $U(x_{min}) = -\infty$. The postulate in question is, I believe, the assumption of continuity, the following version of which is taken from Luce and Raiffa (1957, p. 27).

> Assumption 3 (Continuity). Each prize A_i is indifferent to some lottery ticket involving just A_1 and A_r. That is to say, there exists a number u_i such that A_i is indifferent to $[u_iA_1, 0A_2, \ldots 0A_{r-1}, (1-u_i)A_r] \ldots$.

Their discussion of it is quite appropos and becomes more meaningful in the present context if we read income x_i for prize A_i and $U(x_i)$ for u_i. Let us also suppose (in order to make the argument as forceful as possible) that an income of x_{min} implies death; x_{min} is just below the subsistence level. Now people regularly go to war, fly airplanes, cross streets and a host of other activities to which one could reasonably well associate a finite objective or subjective probability of causing death. Thus, people do indeed behave as if they had established indifferent trade-offs (values of u_i) in situations where one of the possible outcomes would leave them destitute or, what may reasonably be considered even worse, dead.

Having somewhat laboriously established the existence and inadmissibility of infinite values in previously suggested utility functions, we might hope to remove this difficulty while attempting to retain as much as possible of the economic experience which led to the formulation of these functions. It seems reasonable to suppose that the before-mentioned utility functions were predicated on the behavior of people in the upper income brackets for much the same reason that Pareto's law was postulated to graduate incomes in these same brackets, namely, the availability of data. We might then expect that the marginal utility function for which we are searching would asymptotically behave like

$$\frac{dU(x')}{dx'} \rightarrow \frac{k}{(x')^{1+\delta}} \qquad \text{(for large values of } x' = x - x_{min}) \qquad (32)$$

where $\delta > -1$. Such a form would at least generalize three of the four suggestions found in the literature.*

It is instructive to pause here momentarily and consider the utility function that would result if the asymptotic dependence in the above relation were replaced by one of equality, namely,

$$U^A(x') = -\frac{1}{(x')^{\delta}} \quad ; \quad \text{(for } x' > x_{cutoff}) \qquad (33)$$

which would only be considered approximately valid above some ill-defined cutoff value x_{cutoff}. Note that $U^A(x')$ becomes $\log(x')$ when $\delta = 0$ and that for the sake of simplicity I have set the constant of integration equal to zero and $k/\delta = 1$ in accordance with relation (20). Suppose that individuals in the upper income brackets acted as if they were maximizing

---

*Indeed, preliminary fits obtained for 1935-36 U.S. data indicate values of δ which are between that proposed by Bernoulli ($\delta = 0$) and that proposed by Dalton and Jordan ($\delta = 1$).

(33) whenever incomes less than x_{cutoff} were not involved and that the value of δ varied somewhat from individual to individual. Individuals could then be characterized as more or less "conservative" (in the sense introduced by Mosteller and Nogee, p. 375) depending on their value of δ, in that the amount of money which they would consider equivalent to any risky prospect would be a monotonically increasing function of δ. This can be verified by considering the well-known mathematical result (Norris, 1938) that

$$\phi(t) = [\int_0^\infty x^t \, d\psi(x)]^{1/t} \qquad (34)$$

is a monotonically increasing function of t. The function $\phi(t)$ represents a family of "average" values of x which depends on t. For t = 1 this function is the arithmetic mean; for t = 0 it is the geometric mean, etc. For our purposes, the Stieltjes integral involving the arbitrary distribution function $\psi(x)$ can be considered equivalent to the kind of expectation value found in equation (18) which involves the arbitrary prospect $\{p^j(x_i)\}$. We then need only identify $t = -\delta$ and $U^A(x') = -[\phi(t)]^t$ to obtain the desired result. This, in turn, gives us some insight into the significance of δ (which I have dubbed the confidence parameter) at least insofar as upper income groups are concerned.

The simplest way of satisfying (32) while eliminating the objectionable infinity at x_{min} appears to be the introduction of an exponential factor $e^{-b/x'}$, where b is a strength parameter indicative of the income level above which (32) begins to hold. This factor has been further modified to be

$$[\delta b e^{-b/(x')^\delta}]$$

for the somewhat arbitrary reason that

$$\frac{dU}{dx'} = \frac{\delta b e^{-b/(x')^\delta}}{(x')^{1+\delta}} \qquad (35)$$

integrates easily to

$$U(x') = \exp(-b/x'^\delta) \ . \qquad (36)$$

Alternately, equation (36) can be written as

$$U(x') = e^{-1/(x'/c)^\delta} \qquad (37)$$

where $c = (b)^{1/\delta}$. Again for the sake of simplicity, the constant of integration has been set equal to zero and k equal to unity. Note that (b,δ) or alternately (c,δ) are the constants referred to in equation (21). In contrast to the decreasing marginal utilities $[(d^2U/dx^2) < 0]$ characterizing the various forms of equation (32) over the entire range of possible incomes, the functional form in (36) is characterized by increasing marginal utilities $[(d^2U/dx^2) > 0]$ from x_{min} to the inflection point x'_I, where

$$x'_I = c(\delta/1+\delta)^{1/\delta}$$

and decreasing marginal utilities thereafter. Following Friedman and Savage (1948), this property makes possible the explanation (or "rationalization") of many insurance and gambling phenomena. Detailed consideration will be given to equations (21) and (36) in light of this paper and other important contributions to utility theory which postdate *The Theory of Games and Economic Behavior* at a later time. For now it is well to keep in mind that equation (36) has not been *derived* from first principles. The choice of this particular functional form has been motivated (however imprecisely) by observations of economic behavior and, further, such observations (such as the fit to 1935-36 U.S. income data to be presented shortly) provide the basis on which it can be said to be validated. The second lecture will therefore deal with statistical verification and certain theoretical properties of the size distribution of income specified by equations (28) and (37).

PART II

INTRODUCTION

In this second lecture the methodology developed previously will be applied to the determination of the functional form of the size distribution of income. A statistical fit to the 1935-36 United States income data will be made using the theoretical distribution. It will be determined that the agreement is excellent in both the extremes of the distribution, as well as for the intermediate incomes. The practical as well as conceptual difficulties in constructing a definition of income within a society will be discussed with reference to available data. This discussion will provide a framework in which the more normative or interpretive aspects of the results may be addressed. These interpretations determine the degree of correlation between the model and the behavior economists believe is reasonable for members of a rational society. Tentative suggestions are presented for future modifications of the model as indicated by the interpretation of the results.

QUASI-STATIC CHANGES IN p(x)

Having specified the average societal utility function,

$$U(x_i, \delta, c) = e^{-1/(x_i/c)^\delta} \tag{38}$$

the single constraining equation

$$\sum_{i=1}^{n} [U(x_i) - \overline{U(x_i)}]^2 \, p[U(x_i)] = c_1 \tag{39}$$

and associated definition

$$\overline{U(x_i)} = \sum_{i=1}^{n} U(x_i) p[U(x_i)] \tag{40}$$

are sufficient to determine a "best" or most probable size distribution of income

$$p[U(x_i)] = \frac{e^{-(1/2\sigma_U^2)[e^{-1/\{(x_i/c)^\delta\}} - \overline{U(x_i)}]^2}}{\sum\limits_{i=1}^{n} e^{-1/2\sigma_U^2 [e^{-1/\{(x_i/c)^\delta\}} - \overline{U(x_i)}]^2}} \tag{41}$$

via the entropy maximization approach. Let us assume that our origin of measurement is x_{min} and drop the primed (x') notation. If we then go to the continuous limit and accept the assumption described following equation (29), the expression for p(x) becomes

$$p(x)dx = \frac{e^{-(1/2\sigma_U^2)\left[e^{-(1/\{(x/c)^\delta\})}-\overline{U(x)}\right]^2}e^{-(1/\{(x/c)^\delta\})}\left[d(x/c)/(x/c)^{1+\delta}\right]}{\int_0^\infty e^{-(1/2\sigma_U^2)\left[e^{-(1/\{(x/c)^\delta\})}-\overline{U(x)}\right]^2}e^{-(1/\{(x/c)^\delta\})}\left[d(x/c)/(x/c)^{1+\delta}\right]}$$

$$(42)$$

The particular parameterization for $U(x_i)$ given in equation (38) was chosen because p(x) dx can then be written entirely as a function of (x/c). This has certain advantages in terms of scaling properties. Certainly the success with which (42) graduates income data should not depend on the units of money or time in which incomes are expressed. The above form, therefore, allows us to compensate for changes in the designated units of measure by making a corresponding change in what we shall term the scale parameter c. Notice that c and δ parameterize the transformation for x-space to U-space and must be exogeneously specified. Once the value of c_1 is known (at least in principle), then equations (39) and (40) constitute a system of two equations in two unknowns which can be solved for σ_U and $\overline{U}$. As can be seen from equation (41), σ_U and $\overline{U}$ parameterize the probability density in U-space and may thus be further differentiatated from the external parameters c and δ.

Incomes (both real and inflationary) have risen greatly over the last twenty or thirty years while c_1 has remained nearly constant. Consequently, we shall have to allege that one or both of the external parameters (δ,c) has also changed if our density function (42) is to self-consistently graduate income data from year to year. Dynamic changes in income (which inherently imply some disturbance of the equilibrium condition) can indeed be placed in a consistent theoretical context if they are associated with changes in the external parameters. The quasi-static approach necessarily assumes that the time scale on which individuals take actions which might possibly affect their income status* is quite short compared to the time scale characterizing changes in the overall level of real incomes and changes in the value of the dollar. The size distribution of incomes is thought to be infinitesimally close to an equilibrium condition at all times as people readily adjust to temporal changes occurring at the macroeconomic level.

Starting with 1944 the Office of Business Economics of the United States Department of Commerce has periodically published quite detailed (disaggregated) distributions of family personal income by income brackets.** As shown in Table 3 (Budd, p. xiii) these data are often summarized in terms of the percentage of family personal income received by the various quintiles and upper five percent of consumer units. The rationale for presenting the data in this form stems, in part, from the widespread use of Lorentz diagrams, as cumulation of the percentages given below allows an immediate plot of such a diagram. What little variation can be observed in these figures is largely traceable to changes in the level of business activity (Liebenberg and Fitzwilliams, pp. 11-

---

*Income incidence is the technically correct wording here (Kuznets, p. xxxiii).

**See Budd (p. xiii) for a detailed listing of the appropriate references to *United States Survey of Current Business*.

TABLE 3

PERCENT DISTRIBUTION OF FAMILY PERSONAL INCOME BY QUINTILES
AND TOP 5 PERCENT OF CONSUMER UNITS FOR
SELECTED YEARS

Quintiles	1944	1947	1950	1951	1954	1956	1959	1962
Lowest	4.9	5.0	4.8	5.0	4.8	4.8	4.6	4.6
Second	10.9	11.0	10.9	11.3	11.1	11.3	10.9	10.9
Third	16.2	16.0	16.1	16.5	16.4	16.3	16.3	16.3
Fourth	22.2	22.0	22.1	22.3	22.5	22.3	22.6	22.7
Highest	45.8	46.0	46.1	44.9	45.2	45.3	45.6	45.5
Total	100	100	100	100	100	100	100	100
Top 5%	20.7	20.9	21.4	20.7	20.3	20.3	20.3	19.6

12) if indeed it is possible to operationally define and measure changes
of this magnitude (Goldsmith, 1958, pp. 83-91). Since incomes have indeed been generally rising since 1944, the constancy of the figures in
Table 3 implies that the incomes of the various status groups (quintile
groupings) have experienced uniform proportional increases. Graphically
this means that the Lorentz curves for the years following 1944 lie almost
exactly on top of one another (Budd, p. xi).

In terms of the entropy-utility model, uniform proportional income
changes imply that each member of the set $\{x_i\}$ has increased by a factor
of k,

$$\{x_i\} \rightarrow \{x_i'\} = \{kx_i\} \tag{43}$$

leaving $p(x_i) = p(x_i')$. It does *not* mean that all *individual* incomes increased at the same rate. If one considers individual income changes in
terms of a stochastic process (something which I have consistently
avoided), relation (43) implies only that there should result a time-
dependent distribution which admits uniform proportional increases. Note
carefully that if

$$c' = kc \tag{44}$$

and δ is considered constant, then $\{U(x_i)\}$, $p[U(x_i)]$, and c_1 will remain
unchanged. Furthermore, c can be considered a product of two factors

$$c = c_{REAL}\, c_{INDEX}$$

and a proportional change in c broken down into a change in c_{REAL} times
a change in the price index, GNP deflator, or other appropriate factor
which corrects for the value of money. It is then quite natural to associate uniform proportional income changes with a concomitant change in
c and to term this dynamic process the scaling mode of income change. On
a more intuitive level, scaling mode increases in c can be thought to em-
body rising expectations stemming from both real and inflationary sources.
These expectations seem to rise in proportion with income in such a way
that our relative positions (with respect to the relevant economic norm

or U scale) remain unchanged. Thus the scaling mode concretely embodies
the frequently expressed notion that "...within wide limits, the quality
of human experience would be about the same at one income level as at
another if the *relative* positions of persons and classes remained unchanged" (Simons, p. 25).

It is, I think, significant that both the Pareto and lognormal distribution exhibit the same sort of form invariance under scaling mode income changes as does p(x). If we write equation (2) in terms of x' = kx,
we get the same form in terms of x except that

$$A_1' = A_1(k)^{-\nu}$$

Similarly, if equation (3) is written in terms of x' = kx, it reduces to
a lognormal distribution in x except that

$$\mu_{LN}' = \mu_{LN} + \log k \ .$$

Thus A_1 and μ_{LN} have the same significance for the Pareto and lognormal
distributions as c does for p(x). Furthermore, in all three cases, the
parameters indicative of inequality (namely, ν, μ_{LN}, and σ_U) remain unchanged. Graphically, scaling mode income changes imply that if p(x) or
$p_{LN}(x)$ is plotted on logarithmic probability paper (as is done in Figures
1 and 2), the resulting curve will appear vertically translated while
maintaining its relative shape. Similarly, a scaling mode change in A_1
results in a horizontal translation of the curve (line) which results if
the Pareto relationship is plotted on full logarithmic (log-log) paper.

Given the extremely large number of changes in the individual members of $\{x_i\}$ which could possibly be associated with a change in c or δ,
I think it is highly significant that the principle dynamic process observed leaves the entire set $\{U(x_i)\}$ unchanged. I would argue that
people almost always accept the economic framework (and associated value
structure) in which they work and devote their energies to improving their
lot in terms of the opportunities it offers them. Friedman (1962, p. 167)
states that, "No society can be stable unless there is a basic core of
value judgments that are unthinkingly accepted by the great bulk of its
members." I most certainly agree and would intuitively justify the invariance of $\{U(x_i)\}$ in these terms.

Income changes associated with the confidence parameter δ are inherently more difficult to analyze than scaling mode changes because the
former involve changes in $p(U_i)$ and c_1 while the latter do not. A thorough treatment of δ-associated income changes must await further research
done in terms of the more complex model introduced in the next section.
Nevertheless, I will simply state that change in δ are thought to be associated with cyclic or oscillatory income changes which are in turn
closely associated with the overall level of business activity. These
"cyclic mode" changes can be decomposed, at least formally, into contributions arising from changes in the probabilities $\{p(U_i)\}$ and contributions resulting from the relation between $\{x_i\}$ and the invariant set
$\{U_i\}$. The form which these later contributions take can be better understood if we assume that c remains constant while δ changes to δ'. The
invariance of $\{U_i\}$ then implies that

$$(x_i/c)^\delta = (x_i'/c)^{\delta'} \tag{45}$$

which it is advantageous to rewrite as

$$(x_i'/x_i) = (x_i/c)^{(\delta/\delta'-1)} \tag{46}$$

This latter form facilitates comparisons with the scaling mode for which
the comparable quotient (x_i'/x_i) is constant. Let us assume for illus-
trative purposes that the general conficence level is low (δ large) but
rising ($\delta' < \delta$). If any given income x_i is greater than c (which my preliminary fit indicates is somewhat less than the median income), the
final income x_i' will have increased. Furthermore, the greater the difference between x_i and c, the greater the proportionate increase in x_i'.
Conversely, when business begins to slacken off (i.e., when δ increases
as the level of confidence drops) the very highest incomes will suffer
greater proportional decreases than those which are not quite so large.

For high income groups, I believe that effects associated with the
invariance of $\{U_i\}$ will yield the predominant contributions to those
changes in $p(x_i/c)$ which can be associated with changes in δ. Indeed,
Kuznets has empirically observed the sort of cyclic pattern discussed
above utilizing U.S. income tax data. Kuznets finds (p. 62) that,
"...inequality within the top 5 or 7 percent moves with cycles in business activity... In other words, the relative spread within the top 5 or
7 percent becomes wider during expansions and smaller during contractions."
His massive display of empirical findings also points out how far we still
have to go on the theoretical side of the ledger. Indeed, the entropy-
utility framework tells us nothing at all about the timing or magnitude
of changes due to either mode (which are quite evidently going on simultaneously), or whether there is any necessary interrelationship between
them. The introduction of more conventional economic thought may prove
quite useful in this context.

DEFINITIONS, CONCEPTUAL DIFFICULTIES, AND THE CHARACTER OF
AVAILABLE DATA

At this point we can no longer avoid the question of operationally
defining income and income recipient. To claim the society's total income (or product) is allocated among all its members, (or at least among
all its working members), in accordance with certain theoretical constraints is *a priori* both prejudicial and unrealistic. First of all, it
prejudices us toward the historical (income $\equiv$ production) concept of income which led to the formation of the national income and product accounts. The total of personal *producer-contribution* income (Lampman,
1954, p. 252) can be reconciled with the net national product although
its distribution is complicated by the inherent difficulties of imputing
income to individuals which is earned but not distributed. The most important forms which such income takes are corporate tax liabilities and
undistributed corporate profits although the receipts of foundations and
other tax-exempt organizations pose quantitatively less significant but
still less tractible difficulties. The proportionate imputation of corporate tax liabilities to individuals on the basis of dividends received
is open to serious question. Do the stockholders really pay this tax out
of moneys they would otherwise receive as dividends or is the incidence
of the tax shifted forward to the consumer or backward to other factors
of production, mainly labor? In any event, it is apparent that corporate
financial dealings and the government's corporate, capital gains, and personal income tax triumvirate affect size distributions of income in ways
that are almost certainly not accounted for in our simple model.

The *consumer-oriented* concept of income can be defined as

"inflow to the person of changes in ability to consume. It
is made up of currently received net claims, both earned and
non-earned, on the national product, which are judged to be
in such a form...as to be immediately available for consumer
use" (Lampman, 1954, p. 253).

While this definition is closely associated with the individual's standard of living, it does not go as far as Irving Fisher's which equated income with a flow of services and thus excluded claims which are received but not consumed, i.e., savings (Blough and Hewett, pp. 196-197). Starting from the personal sector of the national income accounts, the major changes which must be made to arrive at consumer-oriented income are the addition of transfer payments, both public (relief, social security, etc.), and private, and the subtraction of personal taxes.

R. M. Haig has defined what I term *economic power* income as "the money value of the net accretion to one's economic power between two points of time" (Haig, 1921, p. 7). Bhatia (1970, pp, 364-365) gives a similar definition of an individual's income as "the maximum amount which he can consume during a period without reducing his net worth." Many other variants of this accounting type of definition exist, and all of them basically attempt to reflect an individual's ability to buy and sell in the marketplace or, from the viewpoint of taxation and social justice, his ability to pay taxes. Economic power income includes all consumer-oriented income plus both realized and unrealized (accrued) capital gains. As such, it takes into account changes in relative valuation amount different kinds of assets.

It should then be apparent that the complexity of society's economic and social structure does not inherently lend itself to the definition of an unique concept of income; any such concept and associated measure of inequality must necessarily be defined in terms of the purpose which it is to serve. Nonetheless, producer-contribution income imputed to individuals comes closest to the concept implicitly utilized in our simple model and is theoretically to be preferred if we hope to bridge the gap between the distribution of income by size and by type, i.e., the division between wages and salaries, entrepreneurial receipts, dividends, etc.

The importance of size distributions of producer-contribution income has been stressed by George Garvey who also points out that (Garvey, 1954, p. 250) "members of the labor force, widened to include recipients of property income only (perhaps in excess of some small minimum amount)," are the most appropriate recipient units to utilize in their construction. Those not actively seeking employment are eliminated from consideration while those who are willing and able to work but cannot find jobs are placed in a special zero-earner category. The simple entropy-utility model presented so far does not incorporate such a zero-earner category. It implicitly assumes that all potential income recipients are fully engaged in the economic process. The model cannot therefore deal with the phenomenon of unemployment. Further, real size distributions enumerate part-time and part-period workers, whereas in theory we can only deal with a distribution among (at least "equivalent") full-time workers engaged over the entire income period in question (Garvey, 1952, pp. 40-45).

The entropy-utility model implicitly allows for, although it does not explicitly deal with, the normal variations in skill and earning capacity associated with the various stages of an individual's life cycle (Morgan). I consider these variations to be of secondary importance as determinants of the size distribution of incomes whereas some of the most sophisticated results (Rutherford) of those imbued with the random shock approach curcially depend on the treatment of "birth and death" processes.

Empirical testing must necessarily proceed in terms of currently available size distributions of income, namely those produced by the U.S. Office of Business Economics (the OBE series beginning in 1944) and that produced by their predecessor, the U.S. National Resources Committee (the 1935-1936 NRC distribution). The concept of income utilized in these studies is summarily "the sum of wage and salary receipts (net of social insurance contributions), other labor income, proprietor's and rental

income, dividends, personal interest payments, and transfer payments"
(U.S. O.B.E., 1953, p. 18). This concept most closely approximates that
of consumer-oriented income discussed above. However, personal taxes have
not been subtracted out and much non-market production and all transfer
payments arising in the private sector have not been included, due to a
lack of sufficient data. The evident differences between this concept
and that of producer-contribution income is expected to have the greatest
relative effect on the income shares of the lowest quintile and uppermost
5% of income recipients. Transfer payments are primarily important to
those in the low income tail of the distribution, whereas the receipt of
dividends (by which gross corporate earnings are imputed to individuals)
is highly concentrated in the upper income tail (Goldsmith, 1956, pp.
10-12; Holland, 1962, pp. 38-56).

        The standard OBE and NRC distributions utilize consumer units, i.e.,
civilian noninstitutional families and unattached individuals, as fundamental income recipients. Such a choice makes sense in terms of social
well-being as it is "difficult to evaluate the living standards of an individual member except within the context of the family" (U.S. O.B.E.
1953, pp. 21-22). On the other hand, it introduces an additional set of
conceptual difficulties for the investigator interested in the distribution of producer-contribution income among individual workers and recipients of property income. Will the deletion of consumer units receiving
only transfer payments and their (non)income leave the relative shape of
OBE distributions unchanged? The standard definitions of income and income recipient may, however, partly ameliorate the problem of part-time
workers, which is of particular importance in determining the income share
of the lowest quintile. Two or more part-time workers in a single family
or an older part-time worker who also receives some retirement benefits
may "look like" one equivalent full-time worker. More importantly, the
difference between the number of wage and salary earners accounted for in
adjusted OBE distributions is usually significantly lower (by approximately 5 million persons in 1947) than the total number who are known to
have worked for wages and salaries at some time during the year. The income associated with these individuals is relatively quite small (which
is thought indicative of the part-time and/or part-period nature of their
employment) and is "distributed by family income brackets, at a later
stage in the procedure, proportionately to the aggregate family incomes
that had been estimated for these brackets" (U.S. O.B.E., 1953, pp. 39-
40).

        While it is exceedingly difficult to estimate the extent to which
these and other effects distort the theoretically preferred distribution
at lower and middle income levels, we are somewhat more fortunate when it
comes to upper income groups. Besides presenting his results in terms of
several income variants, Kuznets has explicitly investigated this effect
of arraying given data sets in terms of income per recipient versus income per spending unit reduced to a per capita basis (Kuznets, 1953, pp.
95-110). While these comparisons could only be made in terms of earnings
and showed upper income shares to be larger "in the distribution by income per recipient than by total income per spending unit; ...yet the
differences in upper group shares thus revealed are, on the whole, small
..." (Kuznets, p. 110). These finds are, I think, even more significant
in light of the extreme changes in the personal composition of upper income groups which are affected by changes in the choice of recipient unit.
For example, Kuznets re-ranked the families and single individuals arrayed
in the 1935-36 NRC distribution according to income per spending unit reduced to a per capita basis in order to make these data comparable with
his own series. While this procedure did not drastically change the
shares of the uppermost 5% of recipient units (it went from 26.7% to

30.1%), close to 50% of the individuals who were in the top 5% before the
re-ranking procedure were not in this group thereafter. The re-ranked
upper 5% accounted for something like 10% of the units (families and single individuals) in the original distribution, as middle-income single
individuals displaced large high-income families on a per capita basis
(Goldsmith, 1954, p. 273).

EMPIRICAL TESTING AND SUBSEQUENT MODIFICATION OF THE MODEL

Momentarily disregarding the host of difficulties enumerated in the
previous section (and others not so enumerated), I attempted fitting the
theoretical form given in (41) and (42) to the original* NRC 1935-1936
distribution presented in Table 2. I initially singled out this distribution quite by chance, but it has several features which recommend it
over the later OBE series. First, the NRC distribution covers the very
highest incomes; the OBE distributions do not break down their uppermost
$50,000-and-over bracket. Second, the levels of personal and corporate
taxes were considerably lower during the earlier years. Furthermore, the
relative importance of undistributed corporate profits with respect to
dividends was significantly less in 1935-1936 than during the post-war
period. We should therefore expect any adjustments for the imputations
of gross corporate earnings to be relatively less important for the NRC
distribution than for the later OBE series.

The fitting procedure depended on an advanced guess-and-check methodology which, I think, could be formalized through the medium of Cohen's
(1950) work. I judiciously chose a pair of values for δ and c, transformed the data into U-space, and looked for the emergence of the sym-
metrical bell-shaped plot predicted by (41). The success of this procedure depended critically on my choice of δ and not so much on my choice
of c. For a given value of δ, c was effectively determined as the value
which would make the points closest to the origin (those corresponding to
the lowest income brackets) most nearly conform to the expected normal
shape. Having gotten some reasonable idea of the values which σ_U and $\bar{U}$
would take on, I was able to estimate the degree to which the normal form
was truncated, i.e., the probability weight below U = 0 and above U = 1.
This allowed me to supplement the above procedure by plotting the data on
(cumulative) probability graph paper. This, in turn, allowed a more accurate determination of σ_U and $\bar{U}$. Fortunately, the truncations were relatively unimportant and this iterative, self-consistent process rapidly
converged toward the fit given in Figure 1.

The particular form taken for the average societal utility function,
namely

$$U(x) = e^{-(x/c)^{-\delta}}$$

accounts for the generally inverted S-shaped appearance which seems to
characterize real income distributions when plotted on logarithmic probability paper. Rutherford (1955, p. 278) gives us a number of interesting
examples,** which also lead us to qualify the above generalization when
it comes to the first concave-downward segment extending into the lower-
most quintiles. In some of these examples no data exist in this region

---

*There exist several readjusted versions of this distribution
(Kuznets, p. 442; Goldsmith et al., 1954, p. 4).

**Rutherford uses probit diagrams rather than logarithmic probability
paper. The reader must therefore rotate these diagrams 90° and take their
mirror image (look through the back of the page).

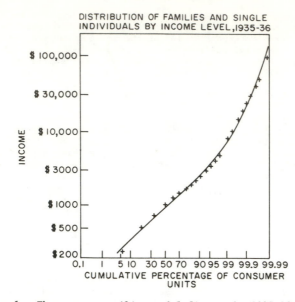

Figure 1.   The entropy-utility model fit to the 1935-36 U.S. income data.
The solid line represents the "best" fit to the form given by
equation (42) with $\delta = 0.58$, $c = 373$, $\sigma_U = 0.16$ and $\overline{U} = 0.38$.

and, even when it does, it sometimes appears to be a linear extension of
the trend in the middle quintiles. It is, I think, noteworthy that if
the average societal utility function were

$$U(x) = 1/x^\delta \quad , \qquad (x > x_{min}, \delta > 0) \quad . \tag{47}$$

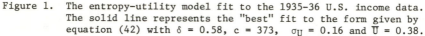

The comparable plot of $P(x)$ on logarithmic probability paper would be
everywhere concave upward and its curvature would increase with the value
of δ.

The best lognormal and Pareto fits to these same data are shown in
Figure 2. As pointed out in the first section, the lognormal distribution fits the data poorly in both upper and lower income tails. The
Pareto law fits quite well in the upper income tail, as indeed the fit
was obtained by considering only the points above $5,000 and then extrap-
olating downwards. Unfortunately, this procedure predicts that everyone
should have had an income above approximately $700, which was quite cer-
tainly not the case. The fit obtained using the entropy-utility distribution is a good bit better than either the Pareto or lognormal fits, but
still leaves something to be desired. The thoeretical curve seems to
meander back and forth through the data points in a systematic fashion
rather than eliciting a pattern of random displacements about a trend
line.

A thorough persual of the methodology used by the National Resources
Committee leads one, all things considered, to a plausible explanation of
at least one of these meanderings. I refer to the rather sharp upward
trend of the data points (at least relative to the theoretically fitted
curve) at approximately the $4,000 to $7,500 level. Indeed, one might
suspect some sort of irregularity in this region simply because it is
here that the two main sources of data (nationwide survey results and

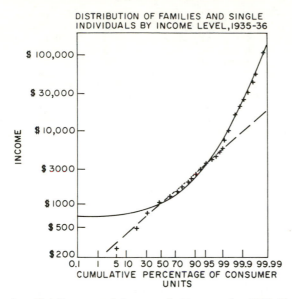

Figure 2.   The Pareto and lognormal fits to the 1935-36 U.S. income data.
The straight dashed line represents the "best" lognormal fit
with $\sigma_{LN} = 0.73$ and $\mu_{LN} = 7.0$.   The solid line represents the
"best" Pareto fit with $\nu = 1.63$.

tabulations of income tax returns) are spliced together after appropriate
adjustments have been made to each.   There seems to be quite general
agreement* that "families with high incomes [are] found to be somewhat
under-represented..." (U.S. NRC, p. 80) by survey data.   It is also generally assumed that non-reporting and under-reporting are most prevalent
in the lowest tax brackets (Natl. Bur. of Econ. Res., 1922 (Vol. 1), p.
130; Kuznets, 1953, p. 447).   Those who compiled the 1935-1936 distribution therefore "decided that the number of families with incomes between
$5,000 and $10,000 [as shown by adjusted tax return data] should be increased by 25 per cent; that the number with incomes between $10,000 and
$15,000 should be increased by 15%; ..." (U.S. NRC, p. 84) and so forth.
Adjustments to take account of similar assumptions regarding under-reporting moved a certain proportion of families from each tax bracket into
the next highest bracket.   This was, in turn, only one of the shifting
procedures which necessitated truncating the adjusted income tax data at
the $7,500 level even though the lower limit for filing during that per-
iod was $5,000.   The final distribution was predicated "on the assumption
that the sample data below $7,500 reflected correctly the relative pro-
portions of families at the different income levels..." (U.S. NRC, p. 86).
I believe this assumption borders on internal inconsistency and that the
NCR estimate of the number (or at least proportion) of people in the
$5,000-7,500 and possibly several adjacent lower brackets should be sig-
nificantly higher than is given in Table 2.
        Not being in a position to readjust the data, I must rely on the
work of Mrs. Goldsmith (1954, p. 4) and her collaborators.   They estimate

*See, for example, Natl. Bur. of Econ. Res., 1922 (Vol. II), p. 417;
Leven, 1934, p. 185; U.S. O.B.E., 1953, p. 28; Kuznets, 1953, p. 442.

that 604,000 rather than 380,266 (the initial NRC estimate) people earned
between $5,000 and $7,500 in 1935-36. I believe this rather large re-
vision supports the methodological critique given above. Unfortunately,
Goldsmith, et al.did not attempt to reanalyze the NRC's treatment of tax
evasion and under-reporting and did not disaggregate their uppermost
$10,000 and over bracket. As a minimal attempt at combining the informa-
tion in the NRC and OBE estimates, I proportioned the figures for the
several more aggregated OBE brackets according to the NRC figures for each
subdivision thereof. This method was perfectly compatiable with the us-
ual OBE practice of proportionally distributing "control totals." The
results are given in the second column of Table 4.

TABLE 4

Income in 10^3 dollars	Number of consumer units	Observed P(x)	Theory P(x)
under .25	1,933,700	5.034	1.028
.25-.50	4,177,400	15.910	8.902
.50-.75	5,256,000	29.594	23.969
.75-1.0	5,350,900	43.525	40.166
1.0-1.25	4,704,400	55.773	53.948
1.25-1.50	3,528,500	64.959	64.612
1.50-1.75	2,723,900	72.051	72.580
1.75-2.0	2,164,200	77.686	78.485
2.0-2.25	1,941,300	82.740	82.881
2.25-2.50	1,428,300	86.458	86.182
2.50-3.0	1,680,400	90.833	90.633
3.0-3.5	1,070,800	93.621	93.345
3.5-4.0	631,200	95.264	95.082
4.0-4.5	395,600	96.294	96.242
4.5-5.0	246,400	96.936	97.048
5.0-7.50	604,000	98.508	98.820
7.5-10.0	231,000	99.1096	99.366
10.-15.	157,880	99.5206	99.720
15.-20.	70,240	99.7035	99.836
20.-25.	41,180	99.8107	99.8891
25.-30.	26,450	99.8796	99.9182
30.-40.	18,570	99.9279	99.9482
40.-50.	8,624	99.9504	99.9831
50-100	13,485	99.9855	99.9865
100-250	4,285	99.99665	99.99647
250-500	947	99.999120	99.99893
500-1000	248	99.999766	99.999774
over 1000	90		

     I have not, however, considered it worthwhile to obtain a theoret-
ical fit to this readjusted data. There exists a very serious positive
discrepancy between the theoretical fitted curve and the data in Table 2
which becomes proportionately larger as we approach the final $1,000,000-
and-over bracket. Figure 1 does not illustrate this divergence at all
well because it is just the beginning at the $100,000 level. The extent
of this divergence is shown in Figure 3 which is a log-log plot of the
uppermost 10% of consumer units versus income. Since the readjusted data
only differ from the original data above $10,000 by a proportionality fac-
tor, this divergence would, by its very nature, be of the same order of
magnitude relative to the appropriate theoretical fit.

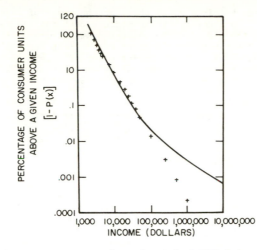

Figure 3.   Pareto representation of original NRC data and first theoret-
           ical fit.

The appropriate form of the initial theoretical fit [1-P(x)] is, as
I have already stated, plotted in Figure 3. It is convex downwards
(toward the origin) and has a much larger curvature than any of the empirical Pareto "lines" to which Macaulay [Natl. Bur. of Econ. Res., 1922
(Vol. II), pp. 344-394] and Johnson (1937) have given detailed consideration. If anything at all, the empirical Pareto lines tend to be slightly
concave downwards in the uppermost income brackets. Indeed, Pareto took
this into account [1964 (Vol. II), p. 305] by means of an exponential
factor in the course of formulating a more general version of his empirical law. In terms of our probabilistic notation, this generalization
can be written as

$$[1 - P(x)] = A_1(x + a)^{-\nu}\ e^{-\beta x} \tag{48}$$

in which a and β are additional positive constraints.

Do people become more adept at understanding their income (legally
or otherwise), and thus avoiding taxes, as their "real" income increases?
Certainly the literature on taxation and the tax laws themselves are re-
plete with mechanisms (personal holding companies, capital gains, Swiss
bank accounts, etc., etc.) which appear to have been designed with this
end in mind (Seltzer, Lundberg). While these mechanisms are quite important in other contexts, I do not believe they hold the key to our present
difficulties. First, it is entirely unclear what "real" income is.
Second, suppose we choose that concept of income (or variant thereof)
which would show the largest percentage increases in the shares of upper
income groups when compared with the standard NRC or OBE concept. Producer-contribution income represents a step in this direction. Its distribution can be approximated by the additional imputation of gross corporate profits to the various income groups in proportion to dividends
received (Holland, pp. 38-56). As mentioned previously, these receipts
are highly concentrated among upper income groups. Furthermore, we might
include realized capital gains (Seltzer, pp. 121-131, 490-496; Liebenberg
and Fitzwilliams) and some rough estimates of tax evasion by means of ex-
pense accounts, depletion allowances, etc. (Lundberg, p. 377). My

calculations were really only a rough check on the order of magnitude of
the figures which such a procedure would yield rather than an estimate of
the type produced by the Office of Business Economics. Nevertheless, the
differences between the theoretically predicted incomes for the very uppermost income groups and the corresponding order of magnitude values
based upon revised concept of income are still so very large as to make
it extremely unlikely that they could ever be accounted for in this
fashion.

This point is reinforced from the theoretical side of the picture
when we note that if $\delta < 1$ (which my preliminary calculations show is
most likely the case), then

$$\sigma^2 = \int_0^\infty (x - \bar{x})^2 p(x) \ dx$$

must necessarily diverge.* Kravis' work (1962, pp. 181-191) shows that
$\sigma/\bar{x}$ can indeed be empirically calculated and tends to maintain a nearly
constant value over a reasonably long period of time. We can formally
avoid this difficulty by dropping the assumption which extended the upper
limit of the x domain from x_{total} to infinity following equation (29).
All the moments of a bounded distribution are necessarily finite. Unfortunately, this formal step will not markedly improve our ability to graduate the 1935-1936 income statistics. However, it does suggest that the
numerical value of x_{total} has a constraining effect on $p(x)$ which our
theoretical framework has not taken into account. Indeed, if we consider
our basic problem to be how a society (of n_{total} people) divides up their
total product (worth x_{total} dollars), then

$$\sum_{i=1}^n n_i = n_{total} \tag{49}$$

$$\sum_{i=1}^n x_i n_i = x_{total} \tag{50}$$

where n is again the total number of income states. Dividing both sides
of equation (49) by n_{total} reproduces the normalizing constraint given in
equation (12). In this context it appears that we should consider the
scarcity constraint

$$\sum_{i=1}^n x_i p'[U(x_i)] = x_{MEAN} \qquad (x_{MEAN} \equiv \text{mean income}) \tag{51}$$

to be just as fundamental as the normalizing constraint. Omission of this
constraint is tantamount to denying that economics has to do with the efficient allocation of scarce resources. In retrospect, the really surprising fact is that the form given by equation (42) fits the data as well
as it does. The initial fit shown in Figures 1 and 3 and the fit to the
revised model in Figures 4 and 5 implicitly testify to the primary importance of the inequality constraint in the context of the present attempt
to explain the distribution of society's total product.

---

*On this score, I can at least claim to be in good company. This
same difficulty is inherent in Mandelbrodt's (1960;61) approach to the
size distribution of incomes which utilizes non-Gaussian stable distributions.

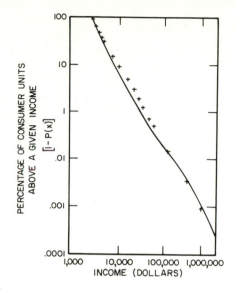

Figure 4.   Pareto representation of revised NRC data and full theoretical fit.

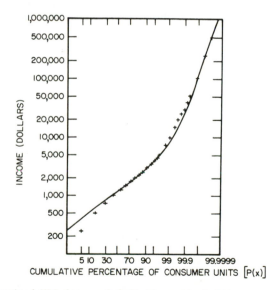

Figure 5.   Revised NRC data and full theoretical fit.

Our model can easily be revised in the following fashion. Let us partition the full $U(x,c,\delta)$ domain (zero to unity) into n equal segments and set up our discrete state space in terms of the U values at the midpoints of the successive segments. Our basic inputs are then the inequality constraint

$$\sum_{i=1}^{n} [U(x_i) - \overline{U(x_i)}]^2 \, p'[U(x_i)] = c_1 \tag{52}$$

and associated definition

$$\overline{U(x_i)} = \sum_{i=1}^{n} U(x_i) p'[U(x_i)] \, , \tag{53}$$

the normalizing constraint

$$\sum_{i=1}^{n} p'[U(x_i)] = 1 \, , \tag{54}$$

and the scarcity constraint

$$\sum_{i=1}^{n} x_i p'[U(x_i)] = x_{MEAN} \, . \tag{55}$$

These in turn lead to the distribution function

$$p'[U(x_i)] = e^{-\lambda_1 - \lambda_1 [U(x_i) - \overline{U(x_i)}]^2 - \lambda_2 x_i}$$

via the entropy approach. It should be noted that λ_1 and λ_2 are not in general independently determined by c_1 and x_{MEAN}, respectively. Thus

$$\lambda_1 = \lambda_1(c_1, x_{MEAN})$$

$$\lambda_2 = \lambda_2(c_1, x_{MEAN}) \tag{56}$$

Going to the continuous limit and writing $p[U(x_i)]$ out in x-space yields

$$p'(x) \, dx = \frac{e^{-c\lambda_2(x/c)} \, e^{-(1/2\sigma_U^2)[e^{-\{1/\delta(x/c)^\delta\}} - \overline{U(x)}]^2} \, e^{-\{1/\delta(x/c)^\delta\}} [d(x/c)/(x/c)^{1+\delta}]}{\int_0^\infty e^{-c\lambda_2(x/c)} \, e^{-(1/2\sigma_U^2)[e^{-\{1/\delta(x/c)^\delta\}} - \overline{U(x)}]^2} \, e^{-\{1/\delta(x/c)^\delta\}} [d(x/c)/(x/c)^{1+\delta}]}. \tag{57}$$

If we once again consider scaling mode income changes [equations (43) and (44)], we see that

$$x'_{MEAN} = k \, x_{MEAN}$$

and that equations (52)-(55) admit decoupled quasi-static solutions such that

$$\lambda'_2 = \lambda_2/k$$

and $\{U(x_i)\}$, $p[U(x_i)]$, and c_1 remain unchanged. This is rather pleasing because it means that the analysis of the scaling mode done in terms of our original model will also hold up in terms of the new model.

Finally, let us consider in what sense the form given in equation (57) approximates the Pareto law for small x. For these incomes, two out of the three exponential factors in the numerator of (57) are nearly constant and we can write

$$[1 - p'(x)] \approx c^{\delta} e^{-\lambda_0} e^{-\{1/2\sigma_U^2\}(1-\overline{U})^2} \int_x^{\infty} e^{-\lambda_2 x} (dx/x^{1+\delta})$$

$$\approx K_1 \int_x^{\infty} e^{-\lambda_2 x} (dx/x^{1+\delta}) \ . \tag{58}$$

Successively integrating (58) by parts yields the following asymptotic power series expansion (Whittaker and Watson, pp. 150-159),

$$[1 - p'(x)] \approx \frac{K_1 e^{-\lambda_2 x}}{\lambda_2 x^{1+\delta}} \left[1 + \frac{1+\delta}{\lambda_2 x} + \frac{(2+\delta)(1+\delta)}{(\lambda_2 x)^2} + \frac{(3+\delta)(2+\delta)(1+\delta)}{(\lambda_2 x)^3} + \cdots \right] .$$

This divergent series will nevertheless closely approximate $[1-p'(x)]$ for very large x even if we keep only the first few terms. Retaining only the very first term yields

$$[1 - p'(x)] \sim \frac{K_1 e^{-\lambda_2 x}}{\lambda_2 x^{1+\delta}} \tag{59}$$

which, upon comparison with equation (48) allows us to tentatively identify

$$\beta \approx \lambda_2 \tag{60}$$

$$\nu \approx 1 + \delta \ . \tag{61}$$

The second of these relations allows us to gain further insight into δ by examining year-by-year tabulations of ν. Each of the values in Table 5 was obtained by plotting the number of tax returns versus U.S. statutory net income on full logarithmic graph paper and calculating the slope of the line connecting the above-$5,000 and above-$500,000 points (Johnson; Tucker, p. 551).

TABLE 5

Year	Pareto slope	Year	Pareto slope	Year	Pareto slope
1914	1.54	1922	1.71	1930	1.62
1915	1.40	1923	1.73	1931	1.71
1916	1.34	1924	1.67	1932	1.76
1917	1.49	1925	1.54	1933	1.70
1918	1.65	1926	1.55	1934	1.77
1919	1.71	1927	1.52	1935	1.75
1920	1.82	1928	1.42	1936	1.72
1921	1.90	1929	1.42		

The value of ν tends to be lower in times of prosperity and higher in times of depression, which is indeed what we would expect if δ is to be consistently interpreted as a confidence parameter.

Further theoretical research and empirical testing are currently being carried out in terms of the modified entropy-utility model.

REFERENCES
1. Aitchison, J. and Brown, J. A. C., *The Lognormal Distribution*
    (Cambridge, 1969).
2. Bernoulli, Daniel, "Exposition of a New Theory of the Measurement of
    Risk" *Econometrica*, 22 (1954). Translated from Latin to English by
    Dr. Louise Sommer from "Specimen Theoriae Novae de Mensura Sortis,"
    Commentarii Academiae Scientiarum Imperialis Petropolitanae V (1738),
    pp. 175-192.
3. Bhatia, Kul B., "Accrued Capital Gains, Personal Income and Saving
    in the United States, 1948-1964," *Review of Income and Wealth*, 16
    (1970), pp. 363-378.
4. Blough, Roy and Hewett, W. W., "Capital Gains in Income Theory and
    Taxation Policy," *Studies in Income and Wealth* (Volume 2), by the
    Conference on Research in National Income and Wealth, a report of
    the National Bureau of Economic Research (New York, 1938).
5. Borch, Karl Henrik, *The Economics of Uncertainty* (Princeton, 1968).
6. Budd, Edward C. (ed.), *Inequality and Poverty* (New York, 1967).
7. Cartter, Allen M., "Income Shares of Upper Income Groups in Great
    Britain and the United States," *American Economic Review*, 44 (1954),
    pp. 875-883.
8. Cohen, A. C., "Estimating the Mean and Variance of Normal Populations
    from Singly Truncated and Doubly Truncated Samples," *Annals of
    Mathematical Statistics*, 21 (1950), pp. 557-569.
9. Dalton, Hugh, *Some Aspects of the Inequality of Incomes in Modern
    Communities*, with an Appendix on "The Measurement of the Inequality
    of Incomes," (London, 1949).
10. Davis, Harold Thayer, *The Theory of Econometrics* (Bloomington,
    Indiana, 1941).
11. Friedman, Milton, with assistance of Rose D. Friedman, *Capitalism
    and Freedom* (Chicago, 1962).
12. Friedman, Milton and Savage, Leonard J., "The Utility Analysis of
    Choices Involving Risk," *Journal of Political Economy*, 56 (1948),
    pp. 279-304.
13. Frisch, Ragnar Anton Kittil, *New Methods of Measuring Marginal
    Utility* (Tubingen, 1932).
14. Garvey, George, "Inequality of Income: Causes and Measurement,"
    Studies in Income and Wealth (Volume 15), by the Conference on Re-
    search in Income and Wealth, a report of the National Bureau of
    Economic Research (New York, 1952).
15. Garvey, George, "Functional Size Distribution of Income and Their
    Meaning," *American Economic Review* (Supplement), 44 (19540, pp. 236-
    253.
16. Goldsmith, Selma and George Jaszi, Hyman Kaitz and Maurice Liebenberg,
    "Size Distribution of Income Since the Mid-Thirties," *The Review of
    Economics and Statistics*, 36 (1954), pp. 1-32.
17. Goldsmith, Selma, "Comment on a Paper by Edward F. Denison," *American
    Economic Review* (Supplement), 44 (1954), pp. 271-273.
18. Goldsmith, Selma, "Income Distribution in the United States, 1952-
    55," *U. S. Survey of Current Business*, 36 (1956), pp. 9-13.
19. Goldsmith, Selma, "The Relation of Census Income Distribution Statis-
    tics to Other Income Data," *An Appraisal of the 1950 Census Income
    Data, Studies in Income and Wealth* (Volume 23), by the Conference on
    Research in Income and Wealth, a report of the National Bureau of
    Economic Research (New York, 1958).
20. Haig, R. M., *The Federal Income Tax* (Columbia University Press,
    1921).
21. Holland, Daniel M., *Dividends Under the Income Tax*, a study by the
    National Bureau of Economic Research (New York, 1962).

22. Jaynes, E. T., "Information Theory and Statistical Mechanics," *Physical Review*, 106, No. 4 (1957), pp. 620-630.
23. Johnson, N. O., "The Pareto Law," *The Review of Economics and Statistics* 19 (1937), pp. 20-26.
24. Jordon, Charles, "On Daniel Bernoulli's 'Moral Expectation' and on a New Conception of Expectation," *American Mathematical Monthly*, 31 (1924), pp. 183-190.
25. Kalecki, M., "On the Gibrat Distribution," *Econometrica*, 13, No. 2 (1945), pp. 161-170.
26. Klein, Lawrence R., *An Introduction to Econometrics* (Englewood Cliffs, New Jersey, 1962).
27. Kravis, Irving B., *The Structure of Income, Some Quantitative Essays* (University of Pennsylvania, 1962).
28. Kuznets, Simon, assisted by Elizabeth Jenks, *Shares of Upper Income Groups in Income and Savings*, National Bureau of Economic Research, Inc. (New York, 1953).
29. Lampman, Robert J., "Recent Changes in Income Inequality Reconsidered," *The American Economic Review*, 44 (1954), pp. 251-268.
30. Leven, Maurice, Harold G. Moulton and Clark Warburton, *America's Capacity to Consume*, The Brookings Institution (Washington, 1934).
31. Liebenberg, Maurice and Fitzwilliams, Jeannette, "Size Distribution of Personal Income, 1957-1960, Role of Capital Gains, Earnings, and Supplementary Incomes," *U.S. Survey of Current Business*, 41 (1961), pp. 11-21.
32. Luce, Tovert Duncan and Raiffa, Howard, *Games and Decisions: Introduction and Critical Survey* (New York, 1957).
33. Lundberg, Ferdinand, *The Rich and the Super-rich: A Study in the Power of Money Today* (New York, 1957).
34. Mandelbrot, Benoit, "Stable Paretian Random Functions and the Multiplicative Variation of Income," *Econometrica*, 29, No. 4 (1961), pp. 517-543.
35. Mandelbrot, Benoit, "The Pareto-Levy Law and the Distribution of Income," *International Economic Review*, 1 (1960), pp. 79-106.
36. Mendershausen, Horst, *Changes in Income Distribution During the Great Depression, Studies in Income and Wealth* (Volume 7), by the Conference on Research in Income and Wealth, a report of the National Bureau of Economic Research (New York, 1946).
37. Menger, K., "Das Unisichesheitsmoment in der Wertlehre," *Zeitschrift fur Nationalokonomie* (1934), pp. 459-485.
38. Morgan, James, "The Anatomy of Income Distribution," *Review of Economics and Statistics*, 44 (1962), pp. 270-283.
39. Mosteller, Frederick and Nogee, Philip, "An Experimental Measurement of Utility," *Journal of Political Economy*, 59 (1951), pp. 371-404.
40. National Bureau of Economic Research, Inc., *Income in the United States, Its Amount and Distribution, 1909-1919*, by the Bureau's staff members: Wesley C. Mitchell, Wilford I. King, Frederick R. Macaulay, and Oswald W. Knanth (New York, 1921-1922).
41. Norris, Nilan, "Inequalities Among Averages," *Annals of Mathematical Statistics*, 6 (1935), pp. 27-29.
42. Pareto, Vilfredo, *Cours d'economie politique*, Nouv. ed. par G. H. Bousquet et G. Busino (Geneve, Droz, 1964).
43. Quensel, C. E., *Inkomstfordelning och skatletryck* (Stockholm, Sveriges Industriforbund, 1944).
44. Rutherford, R. S. G., "Income Distributions: A New Model," *Econometrica*, 23 (1955), pp. 277-294.
45. Savage, Leonard J., *The Foundations of Statistics* (1954).
46. Schultz, T. Paul, "Secular Trends and Cyclical Behavior of Income Distribution in The United States: 1944-1965," *Six Papers on the*

Size Distribution of Wealth and Income, Studies in Income and Wealth (Volume 33), by the Conference on Research in Income and Wealth (Lee Soltow, editor), a report of the National Bureau of Economic Research (New York, 1969).

47. Simmons, Henry C., *Personal Income Taxation* (Chicago, 1938).
48. Seltzer, Lawrence H., *The Nature and Tax Treatment of Capital Gains and Losses*, National Bureau of Economic Research, Inc. (New York, 1951).
49. Tribus, Myron, *Thermostatics and Thermodynamics* (New York, 1961).
50. Tucker, Rufus, "The Distribution of Income Among Income Taxpayers in The United States, 1863–1935," *Quarterly Journal of Economics*, (1938), pp. 547–587.
51. U. S. National Resources Planning Board, Industrial Committee, *Consumer Incomes in The United States: Their Distribution in 1935–36*, National Resources Commiteee (Washington, 1938).
52. U. S. Office of Business Economics, National Income Division, *Income Distribution in the United States by Size, 1944–1958*, prepared by Selma Goldsmith, George Jaszi, Hyman B. Kaitz, and Maurice Liebenberg, a Supplement to the *U. S. Survey of Current Business*, 33 (1953).
53. Vickrey, William, "Measuring Marginal Utility by Reactions to Risk," *Econometrica*, 13 (1945), pp. 319–333.
54. Von Neumann, John and Morgenstern, Oscar, *Theory of Games and Economic Behavior*, 2nd edition (Princeton, 1947).
55. Whittaker, E. T. and Watson, G. N., *A Course of Modern Analysis*, 4th edition (Cambridge, 1963).
56. Wilson, Alan, "Notes on Some Concepts in Social Physics," *Papers, Regional Science Association*, 22 (1969), pp. 159–193.
57. Wilson, Alan, "The Use of the Concept of Entropy in System Modeling," *Operational Research Quarterly*, 21, No. 2 (1970), pp. 247–265.
58. Yntema, Dwight B., "Measures of the Inequality in the Personal Distribution of Wealth or Income," *Journal of the American Statistical Association*, 28 (1933), pp. 423–433.